I0082955

SOCIAL MEDIA AND ORDINARY LIFE

CRITICAL CULTURAL COMMUNICATION

General Editors: Jonathan Gray, Aswin Punathambekar, Adrienne Shaw
Founding Editors: Sarah Banet-Weiser and Kent A. Ono

Social Media and Ordinary Life

Affect, Ethics, and Aspiration in Contemporary China

Cara Wallis

NEW YORK UNIVERSITY PRESS

New York

NEW YORK UNIVERSITY PRESS
New York
www.nyupress.org

© 2025 by New York University
All rights reserved

Library of Congress Cataloging-in-Publication Data

Names: Wallis, Cara, author.
Title: Social media and the ordinary : affect, ethics, and aspiration in contemporary China /
    Cara Wallis.
Description: New York : New York University Press, 2025. | Series: Critical cultural
    communication | Includes bibliographical references and index. | Summary: "Social
    Media and the Ordinary examines how diverse marginalized groups - young creatives,
    rural micro-entrepreneurs, domestic workers, and young feminist activists - engage
    with social media to pursue their dreams and better their lives. Through highlighting
    the affective and ethical dimensions of these processes, the book adds insight into larger
    transformations occurring in China"—Provided by publisher.
Identifiers: LCCN 2024031412 (print) | LCCN 2024031413 (ebook) | ISBN 9781479825035
    (hardback) | ISBN 9781479825066 (paperback) | ISBN 9781479825073 (ebook) |
    ISBN 9781479825080 (ebook other)
Subjects: LCSH: Social media—Economic aspects—China. | Equality—China.
Classification: LCC HM742 .W345 2025 (print) | LCC HM742 (ebook) |
    DDC 302.23/10951—dc23/eng/20250115
LC record available at https://lccn.loc.gov/2024031412
LC ebook record available at https://lccn.loc.gov/2024031413

New York University Press books are printed on acid-free paper, and their binding materials
are chosen for strength and durability. We strive to use environmentally responsible
suppliers and materials to the greatest extent possible in publishing our books.

The manufacturer's authorized representative in the EU for product safety is Mare
Nostrum Group B.V., Mauritskade 21D, 1091 GC Amsterdam, The Netherlands.
Email: gpsr@mare-nostrum.co.uk.

Manufactured in the United States of America

10 9 8 7 6 5 4 3 2 1

Also available as an ebook

CONTENTS

Introduction

Social Media and the Ordinary

Back in June 2014 in Beijing, over two consecutive days, I attended two events that illuminate the desires, discourses, practices, local and global power relations, and contradictions at the heart of this book. On the first day, I was invited by a Chinese colleague to attend an event sponsored by Baidu Baijia, the online news platform of Baidu, China's biggest search engine. Nicholas Negroponte, the MIT tech guru, was the featured speaker of what was called the "The Big Talk," in which he shared his ideas about "making and thinking" and an imminent "post ownership world" with his usual idealism.[1] However, I was more interested in the next speaker, Wang Jianzhou, the former CEO of China Mobile. During his briefer talk, he focused on how the future of mobile lay in rural areas and among the very young and the elderly. He envisioned inexpensive smartphones that would be available to nearly everyone, an "epoch-changing" mobile, and the disappearance of the distinction between a computer and mobile phone.[2] The audience of mostly male Baidu employees seemed enthralled by the speakers' utopian visions— which elided consideration of technology's negative possibilities—and the ballroom of the swanky China World Summit Tower, the venue for the "The Big Talk," was still buzzing with excitement when I left.

The next day I attended a very different event, which took place in a small theater in a residential part of Beijing. It was *The World's Factory*, a 90-minute production made up of film clips, songs, and several vignettes about migrant workers, performed by about a dozen semi-professional actors, using minimal props, costumes, and set pieces.[3] The narratives touched on the hardships of rural life and the exploitation of China's factory workers, who mostly come from the countryside, but made a broader critique about the oppression and marginalization of all of China's rural residents and laborers. There was also a segment dedicated to Karl Marx's proletariat who had engaged in labor struggles during the Industrial Revolution in England.[4] Such comparisons implicitly invoked China's socialist legacy and were meant to inspire hope that China's workers would someday, similarly, gain more rights. The show ended with songs performed by the New Worker Art Troupe, a well-known "new" (migrant) worker band.[5]

I found *The World's Factory* moving although not wholly original, having seen several non-governmental organization (NGO)-sponsored theater productions depicting the struggles of China's migrant population over the years. The audience of mostly scholars, labor activists, and migrant workers, including some of the domestic workers I discuss in chapter 3, was clearly moved as well, as indicated by audible sobs and whispered expressions of agreement during heart-wrenching scenes that called out social injustice.[6] Afterwards, during the Q&A, people praised the production and expressed hope that it could help change poor labor conditions. However, the most revelatory part came while a professor in the audience was passionately making a point about social justice. Suddenly a man at the back of the theater criticized the production, shouting something to the effect of, "With capitalism some people will always face exploitation, lose out, or be left behind altogether!" He then angrily asked why the actors couldn't focus on the positive aspects of China's development. When a few people up front disagreed with him, the man charged down the stairs and lunged at one of the actors on stage. Unfortunately, this incident put an early end to a provocative discussion, and everyone exited the theater. As my companions and I walked away, the man was still arguing with others outside. Later, when I checked my social media feed, I saw pictures with "hearts" and comments about the play, but nothing about the confrontation at the end.

"The Big Talk" and *The World's Factory* cast light on China's disparate social, cultural, economic, and technological realities, and in many ways were worlds apart. The former was indeed about "bigness." Baidu and China Mobile, representative of China's big tech, have helped usher in the nation's digital and mobile revolutions. The Baidu event showcased these achievements and the techno-optimism associated with tech visionaries such as Wang Jianzhou, as well as Huateng "Pony" Ma and Jack Ma, the founders of Tencent and Alibaba, respectively.[7] These men were said to embody the creativity and know-how that many believe will lead China to dominate the twenty-first century. That everyone should emulate them was codified in 2015 in state policy on "mass innovation" and "mass entrepreneurship" (Lindtner 2020; State Council 2015b, 2015c). The same year, China's Internet Plus plan and Made in China 2025 initiative were further articulations of the state's focus on technological, scientific, and manufacturing innovation (State Council 2015a, 2015d).[8] Such policies are instrumental, intended to upgrade the nation and its citizens. However, they are also meant to harness affect, particularly that which is positive, in ways that are both diffused (e.g., the excitement circulating in events like "The Big Talk") and top down (which I discuss in more detail later), for the sake of China's prosperous future.[9] Yet, the government's extensive crackdown on big tech six years later, which began with the suspension of Jack Ma's Ant Group IPO in November 2020, and eventually expanded across the entire tech industry and to seemingly unrelated realms such as fan culture, confirmed that in Xi Jinping's China, no person or entity, no matter how successful, is immune from government censure.[10]

The World's Factory, on the other hand, from its humble production values to its narrative content, was about smallness, exposing the wounds, trauma, and yet hope of China's millions of workers, both those in the factories that produce the products that create the wealth of these giant tech companies and millions of others who perform exploitative and/or undervalued labor in a range of locales. When their struggles appear in the wider public domain, sometimes as a "mass event" (e.g., a short-lived protest often enabled by social media), this might lead to superficial fixes but rarely to meaningful structural change.[11] Still, a focus on large-scale public protest and contention, though important, can also limit understanding of the long-term communicative processes involved in more intimate daily struggles for voice and a better life. These processes often take place in encounters with others and through small practices of self-making, which also are frequently articulated to social media. What the "The Big Talk" and The World's Factory share, then, even as they seemingly exist at two opposite scales of action and with two opposite sets of concerns, are a desire for and an ethical commitment to a better life among people who are separated by background, geographic origin, and life opportunities, yet nonetheless connected.

Situated in these competing and complementary discourses, *Social Media and Ordinary Life: Affect, Ethics, and Aspiration in Contemporary China* is an expansive study of how the daily practices, ethical choices, desires, and emotions of under-resourced, exploited, and/or politically disenfranchised individuals are articulated *through* and *to* social media. I ask: How are state discourses, in particular those related to creativity, entrepreneurship, positivity, and self-optimization, materialized in the everyday lives of those who are marginalized? How is their quest to achieve their goals and aspirations ultimately about the constitution of themselves as ethical subjects? What are the affective dimensions of these processes? How is social media articulated to these processes? Based on ethnographic research conducted offline and online at different periods of time over several years, I answer these questions through four case studies: 1) disadvantaged young creatives, men and women, who have migrated to Beijing from rural areas and who use digital media for informal learning and curating their creative but precarious lives in their quest for personal aesthetics; 2) micro-entrepreneurs in rural Shandong province, especially women whose affective ties and position in the patriarchal family often mask or constrain their use of technology for economic enhancement; 3) domestic workers, all women, positioned in urban homes in Beijing as subordinated, gendered labor, who use social media to build community and construct themselves as ethical, empathic subjects in contrast to dominant discourses that portray them as backward and morally suspect; and 4) young feminists spread across China who are part of a loose online network and who deploy social media to find connection and fight sexism and misogyny, often facing social, and for some, economic and political, marginalization in the process. Through centering the individual, the family, the community, and the network, respectively, each chapter reveals how people's passions, struggles, and everyday ethical choices are mutually constitutive with their engagement with social media in what I call neo/non-liberal China. As I discuss in more detail below, neo/non-liberal China manifests as an intensification of prior discourses regarding self-responsibilization, entrepreneurship, happiness, and harmony along with greater marketization, platformization, datafication, social stratification, and an authoritarian sate that harnesses affect to induce pride and patriotism even as it has ramped up control and repression.[12]

By affect, I mean the inarticulable energies, passions, and forces that move bodies, often understood as the ability to affect and be affected, which is closely connected to the Chinese notion of *gan* (感), referring to "affectivity in both of its tenses—stirring and being stirred" (Sundararajan 2014, 183). Often, "stirring and being stirred" leads to feelings and emotions as well as a desire for expression. I understand such expression as "voice" broadly conceived, drawing upon Bakhtin's (1981) concept of voice as a dialogic process characterized by heteroglossia, which emphasizes a multiplicity of voices and the context and social location of speakers. This book is also informed by Nick Couldry's (2010) notion of "voice as process," or how people understand and express their lives, and "voice as value," which involves valuing these narrative processes and diverse voices.[13] Social media is not merely a vehicle for the expression of voice. It is one element in a broader and shifting assemblage of multiple and contested communicative practices, engagements, bodies, flows of emotion, ideologies, values, discourses, and so on (Savat and Poster 2009; Slack and Wise 2015; Wallis 2013a).[14] Ultimately, in my analysis, the starting point is people rather than technology. I bring their disparate voices together to argue that they should be valued equally.

In recent years, in studies of technology and society in and outside China, there has been much attention given to big data, algorithmic control, surveillance, and platformization. Rather than focus on these daunting technological forces, in this book I center the "small." To be sure, this study could not exist without big tech's imaginary, so uncontested at "The Big Talk," and the massive investments by telecom companies and the Chinese state in producing an array of platforms, apps, and devices to be used for literally everything by everyone. Although so many phenomena in China often seem outsized, my claim is that providing deeply contextual and intimate portraits that center diverse marginalized individuals' passions, feelings, and ordinary ethical judgments, as these are articulated to social media, enables deeper insight into their aspirations, desires, and goals as well as the struggles and obstacles they face in realizing them. Their experiences and ordinary communicative practices also gesture to broader ongoing social transformation in China.

Social Media and Ordinary Life seeks to add to scholarship on the specificities of social media and social life in China while also contributing to broader conversations and analyses that decenter and de-westernize media and communication scholarship (Curran and Park 2000; Waisbord and Mellado 2014; Wang 2011). At the same time, several processes documented here find local iterations across the globe, including the uptake of creativity and innovation by governments for economic productivity; the growth of authoritarianism and populist political trends; belief in the promise of e-commerce to reduce glaring economic inequality; and the spread of globally networked feminist movements.

Early in the research for this book, there was much enthusiasm about the potential of Sina Weibo (now simply called Weibo), "China's Twitter," to transform society through serving as a form of "grassroots media" (*caogen meiti* 草根媒体) (Wang and Yang 2012).[15] I was certainly interested in this possibility, having been in Beijing a couple of years earlier when the 2011 Wenzhou high-speed rail crash had occurred, and netizens used Weibo to expose the government's lies about the accident.[16] During that time, many of my urban Chinese friends expressed hope about how Weibo gave ordinary people a voice to lodge critiques, disseminate information, and hold the Chinese government accountable during what I came to call "spectacular" events (accidents, protests, political scandals, cover-ups, etc.). Much popular and academic work shared this focus (Bandurski 2012; Huang and Sun 2014; Kaiman 2013; Wu 2012; G. Yang 2013). From the start, however, it was clear to me that this discourse obscured the fact that these "ordinary" people were primarily urban and educated.[17] I also knew that the inordinate focus on public, usually one-off events (called internet mass incidents, *wangluo quntixing shijian* 网络群体性事件) potentially hindered deeper understanding of the long-term processes involved in small everyday struggles for voice, empowerment, and agency, processes that have been the focus of my research in China for a number of years (Wallis 2013a, 2013b, 2015b, 2018).

In my encounters with domestic workers and rural residents during that time, most of them did not know what Weibo was, and if they did, they had little interest in it, preferring instead to use QQ (for chat) and Qzone (for social networking).[18] Over the years, as they became more adept at using social media, their usage differed from that of urban elites, as they tended to focus very little, if at all, on debating social issues and instead on seeking and providing emotional support or balancing care work while making sacrifices for the economic well-being of their families. Rural women micro-entrepreneurs and domestic workers might have harbored anger at their unfair treatment by an unscrupulous business associate or an employer, respectively, yet they tended to emphasize the positive. This focus reveals not only the strength of gendered rules regarding social interactions, particularly for rural women who are not supposed to "talk back," but also could be read as adherence to a governing discourse of positivity that has been amplified under Xi Jinping as a means of control and value extraction, as I discuss below. Another interpretation, however, is that by deemphasizing negativity they are following modes of human feeling that undergird interpersonal interactions.

The young feminists are also supposed to be bound by such gendered norms and dominant discourse, but because they do give voice to negative phenomena in China and refuse the role of submissive, gentle women serving men, they are cast as unruly and "cancerous." Much scholarship has focused on the eye-catching street performances and social media campaigns of a group of highly visible young feminist activists (Fincher 2018; Li and Li 2017; Wang and Driscoll 2019) as well as various forms of hashtag activism, such as #MeToo (Han 2021; Zeng 2020). In this book I detail how young feminists, most of whom are out of the public eye, are moved to engage in a range of small practices that ultimately are about constituting themselves as ethical subjects through combating patriarchal structures. Such enduring, and under President Xi Jinping, increasingly retrenched gender inequality was on full display in "The Big Talk" and *The World's Factory*, which highlighted male tech gurus, writers, directors, scholars, singers, and (mostly) male factory workers.

The groups featured in each chapter of this book are diverse, yet there are several threads linking them together. At the macro level, each aligns in some way with key concerns of the party-state in neo/non-liberal China. For example, state policies and official discourse emphasize innovation, e-commerce, and self-improvement to fuel China's rise, and *all* Chinese citizens are meant to respond to such discourse, even if some are seen as more capable than others. Thus, domestic workers and rural residents like those featured in this book, the supposed "vulnerable groups" with low *suzhi* (素质), or "quality" (moral, physical, mental, educational, etc.), are often presented as "problems" to be solved—through, for example, rural economic development, urbanization, and the expansion of opportunities for consumption and leisure that are supposed to make them happy.[19] However, when they do not respond in line with the government's wishes or when they are no longer deemed useful, they are subject to punitive measures. One example is the large-scale, and often cruel, eviction of migrant workers, the so-called "low-end population," from cities like Beijing and the destruction of their homes and places of business and employment that started in early 2017 under Xi Jinping's "Beautification" campaign (see figure I.1) and escalated after several people died in a tragic building fire in a suburb populated with migrant workers.[20]

The young creatives from rural areas exist a bit below the radar of such policies, yet they also are disadvantaged. Like urban residents their age, they belong to the "post-80s" and "post-90s" generations and grew up during the one-child policy and as China embarked on its rapid privatization and marketization and simultaneous dismantling of its collective past.[21] Unlike their urban counterparts, however, many were not singleton children (despite the one-child policy), and they did not have privileged upbringings.[22] Nonetheless, they have similar desires for self-fulfillment in the city, where they face uncertainty. In comparison, most of the young feminists featured in this book were raised in relative privilege because of their urban status, and all benefited by earning bachelor's, and in some cases master's, degrees. They and the young creatives are of the same generation, yet many of the young feminists, as only daughters, did not have to compete with siblings for resources (Fong 2002). However, the feminists' refusal to accept the gender status quo has resulted in their social and political marginalization. In 2015, five young feminist activists, later deemed the "Feminist Five," were arrested and detained for 37 days after planning to hand out stickers to raise awareness about sexual harassment on public transport in cities, including Beijing and Guangzhou, on International Women's Day. Since that time, a form of consumerist, popular feminism (Banet-Weiser 2018) has continued to emerge in urban China even as feminist ideas regarding misogyny and sexism are stigmatized and maligned in the public sphere (Yin 2022).[23] To counter the economic marginalization they face, some feminist activists design and sell feminist clothing

Figure I.1. Concrete over windows in Beijing, November 2017. Photograph by the author.

and accessories online and thereby are somewhat ironically folded into state discourses that entrepreneurship and creativity are key to China's continued prosperity and rise as a global hegemon.

If we turn to the micro level, *The World's Factory* serves as an entry point for mapping the lived connections between these four groups. It featured the plight of rural residents and migrant workers; drew an audience that included domestic workers and feminists (the latter of whom also hold workshops and training sessions for the former); and portrayed issues and themes, such as social justice, that several of the young creatives address in their work and devote their energy and limited resources to promoting. All of the people in this book face different types and degrees of marginalization, and their communicative practices are geared toward opening up spaces for agency, however limited, to pursue their goals. Such goals include crafting a personal aesthetics, ensuring economic security, offering emotional support, and advocating for greater rights and recognition. These are often articulated to social media, as when domestic workers use WeChat (Weixin 微信) to uplift one another, but also through labor and feminist activists who speak on workers' behalf, sometimes in the process overpowering the voices of the marginalized (Gleiss 2014). Moreover, although emerging from very different contexts, my informants' struggles for voice and agency reveal the importance of emotion and affect and the ordinary ethical frameworks that are supported, challenged, and in some cases, violated in these struggles. At different times, people comply with, contest, and quite often, as a strategy of survival, seem to turn a blind eye to governmental regulatory discourses that ultimately are about exerting control and extracting value.

In the remainder of this introduction, I lay the groundwork for the rest of the book by first situating my analysis in a body of scholarly work on ordinary affects and ordinary ethics. I then elaborate on what I call neo/non-liberal China, in order to highlight both continuity and distinctive shifts that have emerged over the last several years. Next, I focus in more detail on an array of processes that are encapsulated in Chinese therapeutic governance and the psychologization of society, which are constitutive of the affective and ethical dimensions of my participants' social media use. I then briefly survey research on social media use in China as well as a body of scholarship that examines the entanglement of affect, emotion, and digital media done outside China to show how this book complements such work. Finally, after a brief note on methods, I provide a chapter overview.

Affect in Ordinary Life

The role of affect and emotion in people's daily interactions and as part of the regulatory political order has been theorized from a number of perspectives.[24] Anthropologist Kathleen Stewart (2007) illuminates the unpredictability and fluidity of these processes through her work on "ordinary affects." She writes, "The ordinary is a shifting assemblage of practices and practical knowledges, a scene of both liveness and exhaustion, a dream of escape or the simple life" (1). She adds that ordinary affects are the "varied, surging capacities to affect and to be affected that give everyday life the quality of a continual motion of relations, scenes, contingencies, and emergence" (1–2). Ordinary affects generate public feelings but also circulate in intimate life. As they are embodied, their intensity can diminish or increase, in the latter case generating thoughts and feelings that could lead to the envisioning of new possibilities.

Stewart's (2007) highlighting of movement, becomings, and intensities captures how the potentiality of affect directs attention to what Seigworth and Gregg (2010, 9) call the "excess . . . the ineffable, as the ongoingness of process. . . . as sticky . . . as immanence of potential (futurity)," and, referencing Baruch Spinoza, the promise of a "not yet."[25] Although the hope is that this "not yet" is something better, this hope can be endlessly deferred, and/or harnessed as a mode of governance and control.[26] As Lauren Berlant (2011) argued, neoliberal capitalism's ruthless disregard for human life is veiled by discourses that individual effort and hard work will always yield a better future, which leads to attachments that do not ultimately bring people happiness, what Berlant called "cruel optimism."[27]

To return briefly to the scenes of the "The Big Talk" and *The World's Factory*, their "varied, surging capacities" generated a range of private feelings and public emotions. In the former, the vibe in the room was one of excitement and pleasure, stemming from technology's promise of a better life. These feelings potentially overpowered any negative sensation associated with the role of big tech companies in perpetuating state repression, censorship, environmental degradation, exploitative labor, online hate and harassment, and what Petit (2015) calls "digital disaffect," or an engaged boredom and detachment one might feel while interacting with digital media.[28] *The World's Factory* also evoked several emotions—sadness, anger, despair, hope, happiness—and these were intensified differently in individual bodies, as manifested through the tears, applause, and laughter of audience members in contrast to the heated confrontation at the end. The defense of the party-state in the final unscripted scene reminds us that the presence of certain marginalized groups historically marked as "other," especially when their presence challenges the dominant order, can engender feelings of disgust, contempt, or hate (Ahmed 2014; Hemmings 2005; Lorde 2007).[29] The angry man's desire for the "positive" also can be read as a manifestation of his interpellation into dominant discourses of happiness and "positive energy" (*zheng nengliang* 正能量), which I discuss in detail below.

The potential of affect as a motivating force and an emergent flow, though sometimes denoted as the "affective turn" in the humanities and social sciences, resonates with feminist work that has long focused on embodiment, the personal, and the power of affect and emotion in constituting subjects, producing social relations, and motivating social change (Abu-Lughod and Lutz 1990; Cvetkovich 2003; Lorde 2007).[30] Although "a basic distinction is that emotion refers to cultural and social expression, whereas affects are of a biological and physiological nature" (Probyn 2005, 11), often the two terms are used interchangeably, particularly by feminist scholars.[31] Because subjective, embodied experience, rather than abstract theoretical distinctions between feelings, emotions, and thoughts, are my concern, I find Sara Ahmed's (2014) focus on the *how* of affect and emotion, or, what emotions *do*, particularly useful.[32] Ahmed turns our attention to the "sociality of emotions" (e.g., emotions are not solely individual but relational), how objects generate (not necessarily shared) feelings as they make an "impression," and how through their circulation they create "affective value," producing attachment and affinity or disagreement and aversion.[33] Like Ahmed and others, I often use "affect" and "emotion" interchangeably. However, for analytical clarity, at times I use the former to denote something inarticulable, or that "registers on the body" (Gould

2009, 19), and the latter as the subjective expression of sadness, anger, joy, etc.[34]

The relational aspects of affect and emotion connect to core understandings of the person and the social in China, where the mind and body are not understood as separate (in terms of a dualism), and bodies are seen as dynamic, made up of vital energy (Sun 2011). Importantly, in dominant Chinese thought, the prominent component of being is the heart (*xin* 心), which is understood as "the seat of cognition, virtue, and bodily sensation," the "origin of all emotions and the grounding space for all aspects of bodily and social well-being" (Yang 2015, 12).[35] The Chinese word for psychology, *xinlixue* (心理学), literally translates as "heart-reason study." However, in contrast to western understandings of psychology, which emphasize the individual's inner psychic life, in China the focus is on the intersubjective (Yang 2014b). Historically and still today, the "individual" does not exist apart from the social relations in which they are embedded (Fei 1992; Hwang 1987; Yang 1994).[36]

Similarly, affect and emotion are not seen as strictly personal but as intersubjective, or constituted in social relations and situations (Kleinman and Kleinman 1991; Yan 1996; Y. Zhang 2007). As mentioned earlier, *gan*, or stirring and being stirred, correlates closely with understandings of affect. Louise Sundararajan (2014) argues that the capacity to be moved (or stirred), namely through having feeling or empathy for another, is more valued than "stirring." She adds that Chinese people believe that emotion (*qing* 情) "discloses something that is true about the person and the world" and that *qing*, rather than distorting reality, grounds people in reality (181). Yanhua Zhang (2007, 55) notes that *qing* "stresses the interaction of human beings with their environment."[37] Several derivations of the word *qing* reveal that feeling and emotion are constitutive of subjectivity and social relations, including *ganqing* (感情), or "sentiments or feelings, emotional attachment," *aiqing* (爱情) or "good feelings and love between couples," and *renqing* (人情), or "social obligation and ethics, a social network" (Y. Zhang 2007, 55), also defined as proper human feelings, modes of social interaction and social obligations, and a resource in social exchanges (Hwang 1987; Yang 1994).[38]

In the last decade, a growing body of critical/cultural scholarship (much of it from anthropology) has joined western theories of affect with deeply rooted Chinese notions of social relationships, mind and body, and spirit/energy. This theoretically rich work has unpacked the role of affect and emotion in modes of subject making, self-cultivation, and effectivity within various cultural sites and social phenomena, including television melodrama (Kong 2014b), state-sponsored re-employment training for laid-off state workers (Yang 2014a, 2015), child-rearing among middle-class urban parents (Kuan 2015), and Chinese migrant workers' social interactions with locals in Africa (Wu 2021).[39] Teresa Kuan (2015, 199) argues, "Both affect theory and indigenous Chinese thought posit a dynamic universe composed of fluid and transformative forces, leading to myriad change and differentiation."[40] I seek to add to such scholarship through examining the entanglement of state discourses and policies—around creativity, entrepreneurship, self-optimization, and positivity—that seek to harness affect to regulate bodies and social media as it is inseparable from feelings of pleasure, pain, hope, and disappointment as well as happiness, sadness, shame, anger, fear, and myriad other emotions. For the participants featured in each chapter, social media was articulated to social (dis)connection, financial livelihoods, the ethical constitution of the self, and efforts at social change, with sometimes contradictory outcomes. Its use could facilitate escape but also feelings of "disalignment" (Döveling, Harju, and Sommer 2018) or exhaustion (Gregg 2011; Petit 2015), a phenomenon I experienced in trying to keep up with social media while writing this book (see figure I.2).[41]

Ethics in Ordinary Life

One day as I was talking with a young creative in Beijing, we discussed what he perceived to be the appropriate manner of posting one's artistic expressions in a WeChat group that he was in—a group that was made up of fellow aspiring artists. He said there were times when he struggled with whether he should post something or not, partly due to insecurity about whether his skill was up to par (if it wasn't, there was a risk he could lose face), but also because he did not want to appear too competent in a way that might make others in the group feel insecure about their own abilities (and then also not post for fear of losing face).[42] His words illuminate how feelings and protecting one's face and the face of others are constitutive of social interaction within the social networks

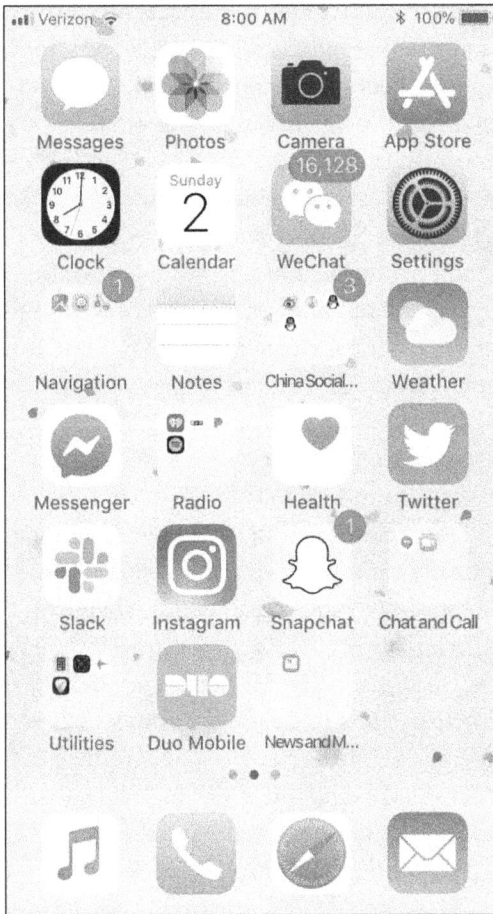

Figure I.2. Too many WeChat messages. Photograph
by the author.

(*guanxiwang* 关系网) of which a person is a member. They also reveal
the connection between emotion and ethical behavior. In other words,
they exemplify the multiple meanings of *renqing* noted above, although
he never used that word in our conversation.

Consideration of one's own conduct, goals, and desires in relation
to others as embedded in one's social media use gestures to the importance of attending to "ordinary ethics," or the ethical dimensions of
practices, decisions, actions, and evaluations that make up daily life and,

thus, selfhood (Lambek 2010; Das 2010, 2012).[43] Scholars who focus on ordinary ethics argue that ethics is "immanent" in human interaction and embedded in the everyday (Lambek 2010, 2015a, 2015b, 2017). This perspective draws attention to judgments about what is right or wrong in particular circumstances instead of universalist notions of right and wrong or good and bad, or to particular rules of a given culture. Das (2012) argues that rather than "orienting oneself to transcendental, objectively agreed upon values," a "descent into the ordinary" focuses on "the labor of bringing about an eventual everyday within the actual everyday" (134). Such a view of ethics connects to both mundane and extraordinary human strivings. Perhaps it goes without saying that these often are articulated to social media.

In much of the work on ordinary ethics, scholars do not make a fine distinction between morals and ethics. If such a distinction is made, ethics points to "the field of action or practical judgment rather than to what is specifically right or good" (Lambek 2010, 9) and morals are viewed as more top-down, structural, and normalizing (Lambek 2010; Mattingly and Throop 2018). Certainly in the Chinese context, morals are often associated with government calls for strengthening "socialist morality" and/or the tenets of Confucianism, which is usually understood to be about rules of behavior.[44] However, many have also argued that ultimately Confucianism is an ethics of practical, situational action (Tu 1985).[45] Research examining ordinary ethics in China highlights how ethical considerations, appraisals, and decisions are part of daily life and can be contradictory (Stafford 2013). A focus on ordinary ethics trains our eye to "the exercise of practical reason as required by the conduct of everyday life," where "ordinary people contend with ethics all the time" (Kuan 2015, 15–16). Again, just as emotions are intersubjective and highly contextual in China, relational, not universalist, ethics guide everyday life. For this reason, I use "ordinary" and "everyday" interchangeably when discussing the ethical aspects of people's social media use, while acknowledging that "ordinary ethics" is a particular theory of ethics emerging from anthropology that has received a fair share of critique, namely that it collapses ethics into the social.[46]

The young artist's honing of his skill, to which social media was intricately connected, as mentioned above, speaks to another aspect of ethics, namely the formation of an ethical self. As I discuss in chapter 1, this mode of working on the self is central in Michel Foucault's later work (1990, 1997) and is a key facet, albeit in a different manner, of Confucianism (Ames 2011; Tu 1985). Although relatively few Chinese people would call themselves Confucians, the rejuvenation of Confucianism by the party-state manifests alongside the diffusion of Confucian, Buddhist, and Daoist modes of self-cultivation in popular culture, which I elaborate on later. This phenomenon aligns with what Kleinman et al. (2011, 10) note as a shift from "collective moral experience of responsibility and self-sacrifice to a more individualistic morality that emphasizes rights and self-cultivation." Thus, even though the young creatives featured in this book engaged in practices that could most overtly be recognized as self-cultivation, a domestic worker's anger at what she perceived as unethical treatment by her employer and her personal use of social media to craft a morally upright self also exemplifies the idea of the ethical as immanent in everyday social interactions. So too does the experience of a young feminist who felt the need to shutter her blog, where she wrote about sexism and misogyny (obviously unethical behaviors), for fear of government retribution.

These examples show how people's pursuit of the self they want to be, and the tactics and strategies they use to get there, are articulated to social media. This claim in and of itself is not new, as the mundaneness of daily life and digital media use has received attention for quite some time from scholars situated in anthropology, media studies, and cultural studies (de Kloet, Chow, and Scheen 2019; Ito, Okabe, and Matsuda 2005). *Social Media and Ordinary Life* complements such work through tracing these processes across diverse marginalized populations and highlighting the nuances of their ethical struggles, passions, disappointments, triumphs, and defeats, as well as, sometimes, their subtle contentment in being.

Neo/non-liberal China

In early 2013 while I was living in Beijing, there was increasing mention, both in official media and in my conversations with friends, of the "Chinese Dream" (*Zhongguomeng* 中国梦), President Xi Jinping's plan for the revitalization—spiritually, morally, economically, militarily, culturally—of the Chinese nation (Kuhn 2013; Xi 2013). Many of my older Chinese friends, who have lived through numerous policy shifts, slogans, and mass campaigns stretching back to the Mao era, wrote off the Chinese Dream as just empty rhetoric. Over the following years, however, the Chinese Dream became ever more prominent in Xi's speeches and in supporting propaganda in the form of news articles, billboards, popular entertainment, school curricula, and political study sessions in state organizations, joining the 12 "Core Socialist Values" as part of the state ideological apparatus (see figure I.3).[47] Xi's "Dream" for strengthening the nation is supposed to ensure the prosperity and happiness of individual Chinese, as long as their pursuit of such aligns with the goals and ideology of the party-state.

As a piece of propaganda, the Chinese Dream is neither catchier nor blander than previous slogans; however, the feelings it is meant to evoke and the economic, social, cultural, technological, and political shifts that have accompanied its rhetoric make it an apt signifier for neo/nonliberal China. This term is meant to capture the contemporary Chinese context marked by the commodification of social life within "market socialism," heightened self-responsibilization, the influence of indigenous Chinese philosophical tenets (from ideas about self-cultivation to deeply engrained patriarchal gender ideologies), the platformization of society, and the rule of an authoritarian state that invokes a therapeutic mode of governance; that is, the state seeks to mobilize "affect and its animated potentialities as objects of governance and sites for value extraction" (Yang 2014b, 12).[48] Neo/non-liberal China is not a break with social formations of previous reform-era regimes (although some call Xi's China "post-reform"). It builds upon prior policies and discourses, for example Xi's desire to fulfill the goal of building a "moderately prosperous society" (*xiaokang shehui* 小康社会), which was first articulated by Deng Xiaoping in the late seventies and was also part of the Hu Jintao-Wen Jiabao-era "harmonious society" discourse. In different ways, it also hearkens back to Mao-era mass mobilizations meant to "raise emotions" (Perry 2002). It amplifies an affective dimension while ramping up surveillance, censorship, and repression,[49] and diffuses throughout society subject positions formerly only available to a relatively small number of people. In theorizing neo/non-liberal China, I build upon prior critical-cultural analyses of contemporary China that have used a Foucauldian lens, including my own, and show

Figures I.3a and b. Posters promoting the Chinese Dream (on the top) and an illuminated display of the 12 Core Socialist Values (on the bottom). Photographs by the author.

continuities as well as changes that have emerged or been intensified under Xi's rule.

At its most basic, the term "neo/non-liberal China" is meant to capture how certain neoliberal economic tenets that were implemented decades ago, including marketization, privatization, closing and/or overhauling numerous state-owned enterprises (SOEs) and industries (which resulted in the layoff of millions of workers), and shrinking the urban social safety net, have occurred in tandem with greater economic inequality and the growth of individual wealth.[50] Commodification of social life and consumption as a way of life are also taken for granted.[51] However, in the mid-2010s, the Chinese economy entered a "new normal," the term Chinese leaders used for a slowing economy and rising unemployment, particularly among college graduates, that came after years of double-digit GDP growth and rising standards of living for the urban middle class. This new normal occurred as the state was seeking to restructure the economy in order to move it away from export processing (e.g., the "world's factory") and low-wage work to high tech and innovation.[52] Even as the state sector was being strengthened, these policies were also accompanied by frequent state discourse on the role of ordinary people in these processes. "Mass entrepreneurship and innovation serves to unleash people's ingenuity and power," proclaimed Li Keqiang at Davos in 2015 (Li 2015). The message was that *everyone* should therefore make, create, and "upgrade" their capacities in order to upgrade the nation (Lindtner 2020).[53] This focus on the necessity to increase human capital for the sake of national progress was not new; rather, it was intensified.[54] Although Li Keqiang's buzzwords faded from state media within a few years, the emphasis on high tech and innovation as key to China's future has remained.

Overlapping with the economic realm, neo/non-liberal China also signals continuities and transformations in social and cultural domains, for example in gender ideologies (which I discuss in more detail in chapter 4). A hallmark of the early reform period was how quickly Mao-era policies to promote "equality between men and women" (*nannüpingdeng* 男女平等), which is enshrined in China's constitution, were overshadowed by state and popular discourses that argued for men and women's natural biological differences. This latter ideology was bolstered in China's commercial culture, through sexualizing and commodifying women's bodies, and it manifested in education, employment, and politics as well as in the private realm (where it never actually disappeared in rural areas) (Evans 1997, 2008; Fincher 2023; He and Zhou 2018; Wallis 2015b; Yang 1999). Since coming to power, Xi Jinping has been very vocal about the importance of "traditional Chinese virtues," elevating Confucian gender roles to emphasize in particular women's "unique role" in raising children and maintaining family harmony.[55] After the one-child policy was dropped in 2015, due to rising alarm over the demographic crisis it brought, more overt state pronouncements concerning women's duty to uphold family values, even including such language in a revised gender law (Xiao 2022), have appeared. Although such traditional ideologies are seen in the commercial realm through, for example, "Ladies" academies that emphasize female virtue and purity (Carrico 2016),[56] many young women seem to be ignoring this call.

The refusal of traditional gender roles is but one indication of the rise of individualism in Chinese society since the onset of the reforms, particularly after marketization and privatization went into high gear in the wake of Deng Xiaoping's 1992 Southern Tour.[57] To Kleinman et al. (2011), for some this meant a "divided self," for example, between the emerging focus on self-interest and former emphasis on the family and the collective.[58] Indeed, the new lifestyles, modes of consumption, and self-styling that accompanied these shifts were thought to have brought ordinary citizens more "micro-freedoms" in daily life even as the authoritarian state enacted certain constraints (e.g., on political rights). To add insight into the massive transformations that occurred during this period and into the first decade of the 2000s, several scholars adopted a Foucauldian lens, in particular engaging with Foucault's (1991) writing on governmentality, or the "conduct of conducts" (2000, 341), to theorize this context, in which politics was seen to have taken a relative backseat to economic development and the emerging arenas for people to pursue diverse aspirations and identities.[59] They noted how individuals from all walks of life were encouraged to self-optimize and become self-reliant "entrepreneurs of the self" through improving their bodies, minds, and spirits "on their own" and in accordance with the state's political limits and numerous methods to try to shape social subjectivities. Ong and Zhang (2008, 2–3) called this assemblage "socialism from afar," meaning that people could "exercise a multitude of private choices" while the authoritarian state continued "to regulate from a distance the fullest expression of self-interest." Other scholars, also trying to conceptualize this milieu in which individuals were encouraged to be relatively autonomous in certain domains (social and economic) but not in others (political), variously termed it "China's governmentalities" (Jeffreys and Sigley 2009), "desiring China" (Rofel 2007), "late-socialist neoliberalism" (Hoffman 2010), and "Leninist neoliberalism" (Greenhalgh and Winckler 2005).[60]

In neo/non-liberal China, in many ways the forces of both self-responsibilization and authoritarian control have intensified. Regarding the former, throughout the reform period the ideal industrious, entrepreneurial self has been prominent in Party propaganda and reemployment training programs that taught the "stagnant" laid-off state worker new skills, such as domestic work (for women) or taxi driving (for men), along with a new "mindset" (Yan 2008; Yang 2015). Jie Yang (2015) has shown how officials and state representatives involved in such programs invoke affect as a means of ensuring social stability through trying to channel workers' "positive potential" (*qianli* 潜力) into various forms of low-wage labor and to suppress their "negative potential" (*yinhuan* 隐患), seen as full of danger and hidden risks. However, in recent years there has appeared a qualitative shift in state discourse and policy regarding how the ideal subject should construct themself. If previous laid-off workers were supposed to become "entrepreneurs" through doing any sort of work other than what they used to do, and if the young migrant women who participated in my research over a decade ago were compelled to be metaphorical entrepreneurs of the self, through improving their *suzhi*, now all Chinese individuals are meant to heed a slightly different call. That is, realizing the Chinese Dream does not only entail structural transformations related to China's marketization of the economy that were previously mobilized to transform people's psyche (Yan 2008). A newer yet related discourse is that *all* people should construct themselves as *innovative* in some way, even though for many the structures and opportunities haven't changed. Again, this call mobilizes positive affect to extract value from human capacities and to exercise political control. It also aligns with both Confucian and pop culture notions of self-cultivation. Again, these latter notions are not new, but they have become more widely diffused.

As just one example of this entanglement of affect, innovation, and markets, for years rural residents, such as the micro-entrepreneurs featured in chapter 2, have been encouraged to engage in e-commerce as a means of economic empowerment, which by extension should bring happiness.[61] During and increasingly since the end of the government's "Zero-COVID" policy, e-commerce has been seen to hold even more promise if it is accompanied by livestreaming. For quite a while, the pillar of success to which rural micro-entrepreneurs were supposed to aspire have been "Taobao Villages," or villages that have gotten wealthy through selling products on Taobao, Alibaba's e-commerce platform. However, this digital entrepreneurial discourse is complicated by the fact that Taobao Villages are relatively few compared to the number of rural villages in China and are located in areas that already had the necessary infrastructure in place (Li 2017), which is certainly not the case for impoverished villages that nonetheless are supposed to heed this call to innovate, create, and by extension, prosper. In the villages where I conducted fieldwork, which were comparatively "wealthy," there still was not the requisite infrastructure even as late as 2022. Moreover, many rural women's primary concerns were how to successfully integrate economic and familial obligations, or how to blend their productive labor with unremunerated care work. Everyday ethical choices then emerge, often articulated to technology, regarding how to build both the moral family and economic wealth. The young creatives I discuss in chapter 1 are also implicated in such discourses even as they see themselves as different from the mainstream precisely *because* they view their passion for creative pursuits as something they were born with and not coming from a state discourse (although in some cases they benefit from this discourse). Thus, the economization of social life, in which all activities are cast in market language and people must always consider their present and future value, is also part of neo/non-liberal China (Brown 2015; Feher 2018).[62]

In the political realm, neo/non-liberal China draws attention to the increased authoritarian control, often deployed through a therapeutic mode of governance, under Xi Jinping, who ascended to the highest ranks of China's leadership when he became General Secretary of the Chinese Communist Party (CCP) and Chairman of the Party's Central Military Commission in November 2012 and President in 2013. Since that time, Xi has accumulated more power and exerted ever greater control in all aspects of everyday life than any leader since Mao Zedong.[63] In 2017, "Xi Jinping Thought on Socialism with Chinese Characteristics for a New Era" was enshrined as Party doctrine into the Constitution.[64] In 2018, Xi abolished term limits (leaders are supposed to serve two five-year terms), effectively making himself ruler for life.[65]

Along with these bold political maneuverings, Xi has ramped up ideological control in ways that both seek to coerce and encourage "proper" behavior. Examples include the teaching of Xi's thought, which the CCP has disseminated through university academic departments, school curriculum, work study groups, television shows, propaganda posters, and even an app.[66] There have also been an endless series of government directives aimed at censoring information and practically eliminating China's nascent civil society, including the quashing of NGOs and persecuting and (ever more frequently) jailing anyone who does not submit to state authority, such as young feminist activists that are part of the network discussed in chapter 4, along with human rights lawyers, whistleblowers, and political dissidents. Fu and Distelhorst (2018) argue that in contrast to prior "fragmented repression" that tended to be sporadic and reactive during Hu Jintao's rule, under Xi Jinping repression has become consolidated and proactive. Furthermore, most forms of activism are framed as a matter of national security and are criminalized.[67] Crackdowns on popular culture—various celebrities, TV shows, livestreamers, etc.—and big tech are an extension of such control and are interrelated (Wang and Bao 2023). China's social credit system is another iteration, blending technological and disciplinary power. Although it is often constructed as a form of Orwellian totalitarian control by western media, it has been unevenly implemented (focusing more on businesses) by different corporate and state entities (Creemers 2018). It too has an affective element that should be underscored: in some cases, public shaming for "bad" behaviors and instilling of pride for "good" behaviors (Liang and Chen 2022). It also enjoys high levels of approval by many citizens (Kostka 2019).[68]

In spite of his heavy hand, Xi has simultaneously fostered an emotional dimension to his leadership that surpasses that of his immediate predecessors. For example, for a number of years, Party propaganda organs promulgated use of the nickname "Xi Dada" (meaning "Daddy Xi" or "Uncle Xi") when referring to the President.[69] Shortly after becoming China's top leader, Xi began calling on media and popular culture to have more "positive energy" (*zheng nengliang*), defined as attitudes, actions, and emotions that are geared towards what is uplifting, optimistic, and hopeful (Hird 2018).[70] The phrase had been used minimally in popular culture (mostly emanating from Hong Kong), yet the state coopted it to attach these positive values and feelings to Party ideology and as a means of "guiding" public opinion, especially online discourse. Yang and Tang (2018) also argue that its sentiment aligned with pop psychology discourses on happiness and positivity already circulating in popular culture and appropriated by previous regimes (as I discuss below).[71]

It is tempting to view *zheng nengliang* as just another empty slogan, yet it became hugely popular and was the top catchphrase online in 2012 (Sun 2012). When I was in Beijing in early 2013, and positive energy had already achieved its meteoric ascent into popular discourse, several of the domestic workers featured in chapter 3 said they intentionally sought to make their social media posts uplifting and affirming, a sentiment that continued over the years.[72] As I argue in chapter 3, their focus on positivity and uplift was not because they uncritically adhered to governing discourses; rather, it was a strategic mode of survival. Part of the reason the feminists that I discuss in chapter 4 are harassed and demonized is that in using social media to draw attention to "negative" phenomena like sexual assault, sexual harassment, and domestic abuse, they counter the positivity that is supposed to be part of the Chinese Dream. Indeed, "positive energy" has joined previous state discourses such as "harmony" (*hexie* 和谐) and "civility" (*wenming* 文明) as a mechanism of control and regulation.

In the neo/non-liberal China assemblage, a final but not determining feature, which has already threaded through this discussion, is the technological realm, which includes the entanglement of informatization, platformization, and datafication. Informatization, or building information communication technology infrastructure, spreading access, and designing applications and services, has been a government priority for years (Hong 2017; Qiang et al. 2009). It is the reason for the near saturation of the internet, smartphones, and various types of digital media into every realm of economic, social, cultural, and political life, and it continues to be a priority. The government's 2021–2025 14th

Five-Year Plan for National Informatization stressed the need for further digital infrastructure and governance systems and the transformation of traditional industries into digital ones.[73]

In its current iteration, informatization is closely linked with platformization. In Van Dijck, Poell, and De Waal's (2018) formulation, the platform society is one in which economics, governance, myriad institutions, and democratic processes are reconfigured and disrupted by the outsized role that online platforms have. Such transformations have occurred in western contexts because platforms themselves are said to be nearly equal in power to governments. In China, since their earliest incarnations, the internet and digital media have been tightly controlled and regulated (Meng 2018; Yang 2009). Moreover, a state-corporate model prevails, as many internet and social media companies are (by design and out of necessity) in a symbiotic relationship with the party-state (Hong 2017; Miao, Jiang, and Pang 2021). Therefore, as de Kloet et al. (2019) have argued, China presents a model to challenge common assumptions about platformization because of the role of the state in platform infrastructure and governance.[74] As already mentioned several times, in China, platforms have become increasingly regulated as the government seeks to create a "positive," "civil," and "harmonious" online environment (Yang 2018). Still, for the people featured in this book, various platforms enabled modes of sociality, business, creativity, activism, and the like, attesting to the diversity of practices that de Kloet et al. (2019) also include in their analysis of platformization in China.

In and outside of China, datafication draws attention to how it is now possible that nearly every action is captured, tracked, sorted, and quantified as people engage with their phones, tablets, and computers, use various platforms, and become the subject of other technologies, including facial recognition and surveillance cameras (Bauman and Lyon 2013). Although outside China, state surveillance is often associated with the social credit system, the Chinese government, in partnership with a large number of private firms, has built a multi-faceted surveillance state (Cheung and Chen 2022; Chin and Lin 2022; Huang and Tsai 2022; Liang, Chen, and Zhao 2021). Private companies as well collect massive amounts of user data. For example, Chen and Qiu (2019) call Didi Chuxing, a popular ride-hailing app, a "data utility."

To sum up, the "neo" in neo/non-liberal is meant to capture the integration of *some* aspects of neoliberal governance—marketization, self-optimization—although China is not neoliberal. In the same way, the "non" is *not* meant to imply that there is no space for people's pursuit of individual goals and desires, including some of which the state disapproves, yet some of the micro-freedoms enjoyed in certain realms in prior decades seemed to have become nano-freedoms, in particular the ability to openly express critique of the party-state without severe sanction. Still, people's passions will always drive them, and creative agency versus domination is not a zero-sum game. Moreover, the above discussion underscores how affect, or the inarticulable energies, feelings, and sensations that propel movement, intersects with contemporary governance in China. Chinese state power can certainly be coercive, and in fact it has become more merciless and pervasive under Xi (rather than governing "from afar"), generating fear, anger, and/or despair for some. However, the party-state simultaneously disseminates feelings of positivity and harmony that summon individuals. These modes of governance intersect with, and are inseparable from, ordinary lived practices and processes that speak to the affective and ethical construction of the self (Yang 2015; Zhang 2015).

Psychologization and the Ethical Care of the Self

The ubiquity and longevity of the discourse of "positive energy," as well as its articulation to the Chinese Dream, are representative of the therapeutic mode of governance that seeks to mobilize affect to encourage a certain mindset and behaviors designed to maintain social stability and bring individual happiness in neo/non-liberal China. It is also connected to a "happiness craze" (Yang 2015; Zhang 2015), which is not just a Chinese version of the "happiness turn" in western neoliberal societies, where popular culture, corporations, and state entities emphasize the cultivation of mindfulness and contentment rather than mental disorder and lack as a means of deflecting attention away from larger structural problems.[75] Rather, happiness as part of a ruling technology connects to a long cultural history and beliefs about what brings fulfillment and a meaningful life. Confucian notions of self-cultivation and proper moral behavior, or what makes one human, are ultimately meant to lead to happiness, which is situated not in the individual but in the collective, through fulfilling filial duties (*li* 礼) to family (Tu 1985). Mao Zedong's quest to build a "happy and prosperous socialist country" had continuity with this past, but happiness was part of an overarching political project (Wielander 2018; Yang 2015).[76] Throughout the reform era, government leaders have also deployed "happiness" in everything from poverty relief programs (Wang 2012) to one-child policy propaganda. The Hu Jintao-Wen Jiabao (2002–2012) discourse of the "harmonious society," which drew upon Confucianism, aimed to increase people's happiness by raising their standard of living, and Xi Jinping has continued to deploy Confucian tenets as part of his Chinese Dream and his effort to elevate "Chinese" (as opposed to western) values.

Despite the emphasis on happiness and the accumulation of material goods, Chinese people have not necessarily been "happy," due to anxiety and discontent caused by glaring economic inequality, perceived social decay, official corruption, and more recently, COVID-19 lockdowns and increasing state control in everyday life. Thus, for years, Chinese academic centers and government think tanks have distributed surveys and conducted research aimed at figuring out what makes Chinese people happy (Wielander 2018; Yang 2015; Zhang 2015).[77] As a mode of affective "kindly power," the state has used distinct techniques to govern certain populations—laid-off urban factory workers are encouraged to attend reemployment training that focuses on "unknotting the heart" and releasing their "positive potential" (Yang 2015); youth in the countryside are given Confucian education in rural boarding schools so that they learn respect and interpersonal ethics; and urban elites are taught Confucian morality as preparation for future leadership (Y. Zhang 2018).[78] All of these efforts are ultimately supposed to make people happy, yet their primary purpose is to ensure social stability and economic development.

Although the government has rehabilitated Confucianism and promotes harmony and positivity, these are not solely top-down processes. They must be situated within an ongoing "psycho-boom" (Kleinman 2010), and what Yang (2014b, 2015) calls the psychologization of Chinese society, which has occurred over the last three decades. These phenomena are manifest in the commercial realm (which is not entirely separate from the state), including call-in radio shows (Erwin 2000), television dramas (Kong 2014b), variety shows (e.g., 2007's "Happy Voice Boy" *Kuaile Nansheng*), and lecture programs, such as Yu Dan's CCTV series *Confucius for the Heart* (Y. Zhang 2018). There is also an individualistic search for meaning, primarily on the part of the urban middle class (Kleinman et al. 2011), which has resulted in a thriving market for pop psychology literature and the formation of study groups whose members read and discuss this literature in their quest for meaning and happiness (Yi 2019). In addition to Confucianism, Chinese Buddhism and (to a lesser extent) Daoism are thrown into the mix for a blending of positive psychology with Chinese philosophical traditions (Y. Zhang 2018; Zhang 2015, 2017, 2020).

Part of the reason so many people are drawn to pop psychology and indigenous spirituality is that there is a widespread sense that their lack of happiness is connected to a moral vacuum in society. For many, faith in the CCP has waned, even though membership is still seen as important for economic advancement and cultivating social networks. Similarly, official rhetoric about "socialist morality" can ring hollow, particularly because it is often used in a regulatory way to denote behaviors—patriotism, altruism, filial piety—that are desired because they will ultimately buttress the authority of the party-state and ensure social stability. Numerous media stories of corruption, violent crimes, or people who stand by and do not help an injured person—exemplified quite graphically in a widely shared viral video of a small child whom no one rescued after she had been run over by a car—also reinforce a belief that traditional values have faded despite all attempts to revive them.[79] Nonetheless, the legacy of socialist ethics, based on notions of equality and putting the collective over the individual, still resonates, in particular because these ideas continue to be deployed in official discourse. The disenfranchised then call upon them in their time of need (Zhao 2008).[80] However, in many ways this is China's own version of cruel optimism, given that self-responsibilization, encapsulated in the *suzhi* discourse, and economization, where people must make themselves valuable, prevail in neo/non-liberal China.[81] This situation seems to confirm Yan Yunxiang's (2010, 2013, 2017) earlier arguments that the contemporary focus on self-development, competition, and individual desires has led to selfishness, greed, and disregard for others.[82] However, Yan (2020) has also argued that greater diversity of lifestyles and values has reconfigured morality in China in positive ways. Nonetheless, the widespread perception of a crisis has led many intellectuals to try to construct a new social ethics (Ci 2014; He 2015).

These tensions were evident throughout my fieldwork, as was a feeling that, quite often, as people navigated choices and challenges, they were left to their own devices to figure out their own path. For example, as I discuss in chapter 3, a domestic worker I knew wrestled with whether to report an exploitative employer to the domestic worker service company that had placed her in the employer's home. In the end, she vented about her situation to the sympathetic ears of other domestic workers during a Saturday drama club practice. Such venting, and the reenactment of trauma through performance as seen in *The World's Factory*, serves as a form of emotional catharsis and aligns with the psychologization of society. After practice, the domestic worker posted a short message of protest on her WeChat, contrasting her ethical and empathic behavior with that of the company and her employer. Social media thus becomes inseparable from the construction of one's current and future ethical self, as affect flows through modes of "self-appreciation," in both senses of the term; that is, displaying and valuing the self, but also accumulating value through, for example, "likes," which ostensibly demonstrate that one has value.[83]

China's Social Media Landscape

Earlier, I discussed the hope many people had had that Weibo, as "grassroots media," would play a role in counter-hegemonic struggles in China. For a time, this discursive public sphere was enabled by Weibo "Big Vs" (users who are verified and have over one million "fans," or followers) and "public opinion leaders" (intellectuals, news anchors, celebrities, star athletes, etc., who are often Big Vs) (Svensson 2014; Tong and Lei 2013; Wang and Yang 2012; G. Yang 2013).[84] In using this medium, these public opinion leaders and NGOs often took it upon themselves to speak on behalf of the marginalized (Gleiss 2015; Svensson 2014), whose own voices were (and are) largely left out of such conversations (Liu and Tian 2007; P. Zhang 2013). However, after a series of government crackdowns, Weibo's influence declined, at least for a while.[85] In its place, WeChat, originally introduced by Sina rival Tencent in January 2011 as a mobile instant messaging (MIM) app, quickly developed into a mobile app for nearly everything used by everyone.[86] Because rural residents and migrant workers were already the primary users of Tencent's QQ and Qzone, it was relatively easy for them to shift from these platforms to WeChat after Tencent enabled this functionality.[87] Aside from its chat/video chat components, and its "Moments" (*pengyouquan* 朋友圈, or literally "friends circle" in Chinese), which is like Facebook's "wall," WeChat includes a "wallet" and "mini programs" for all kinds of functions (Chen, Mao, and Qiu 2018).[88] I believe that this shift from social media being seen as something that could "change" China to something essential for life is aptly captured in the waning of the term "grassroots media" and the prominence instead of "self media" (*zimeiti* 自媒体).

Although WeChat seems to be used by people from all segments of Chinese society, the platform economy in China is often stratified according to class-based and rural/urban distinctions. The aforementioned Taobao Villages are one manifestation of this phenomenon, but it is seen in other realms as well. For example, numerous platforms provide video streaming and livestreaming of all kinds of content (e-commerce, gaming, education, concerts, sports, etc.), which has given rise to the *wanghong* (网红), or internet celebrity, phenomenon.[89] Rural residents and migrant workers, including some of the women featured in chapters 2 and 3, are primary users of Kuaishou, whose stars are often migrant workers and marginalized rural youth, who upload music, short humorous videos, and other grassroots content to the platform to try to gain fame and escape poverty (Hou 2020; Li, Tan, and Yang 2020; Lin and de Kloet 2019). However, Kuaishou's content became more sanitized after the platform was disciplined by the government for its "vulgarity," which did not accord with positive energy dictates. Individual platforms, following state regulations, also discipline users. A young feminist I discuss in chapter 4 used to livestream on Bilibili (akin to YouTube), yet she eventually ended her livestream due to harassment and censorship, from viewers and from the platform.

These examples reveal not only how social media reinforces class differences, but also how, as social media has rapidly diffused, the government has tried to balance its dual desires to capitalize on and control its content. From its inception, Chinese social media has been heavily regulated and easily manipulated by the state in the same way that earlier internet communication was and still is (Bamman, O'Connor, and Smith 2012; Jiang 2012).[90] However, although King, Pan, and Roberts (2013) previously found that the state allowed a large amount of criticism and discussion online as long as there were no calls for collective action, the space for such critique has shrunk. Just as ramped-up ideological control through formal institutions is one feature of neo/non-liberal China, the current leadership has taken increasingly aggressive measures to rein in the online realm even as, and precisely because, more and more of everyday life is lived through engagement with platforms.[91] There is now a dizzying array of entities to create and enforce new rules and regulations.[92] This "micro social management" has in many ways created the "positive" online environment the government desires while sacrificing real debate (Hu and Chen 2017; Yang 2018). As well, virulent nationalism, trolling, and "flooding" the online space with distracting messages are allowed for certain periods of time to stamp out dissenting voices (Han 2018; Roberts 2018).

Despite numerous crackdowns and top-down control, Chinese people from all realms have integrated social media into daily life in accordance with their own needs. Several books have traced this development, from the age of BBS to social media as life necessity (Guo 2020; Negro 2017), or have focused on the affordances of a particular platform, such as WeChat (Chen, Mao, and Qiu 2018) or Weibo (Han 2016). Still others have provided a deep portrait of a certain phenomenon, such as online fandom, which offers the potential foundation for a new, networked public sphere (Zhang 2016). There are also rich collections of diverse users and practices (DeLisle, Goldstein, and Yang 2016; Kent, Ellis, and Xu 2018).[93] In each chapter of this book, I engage with research on the social media use of the group that is the focus of the chapter.

Social Media and Ordinary Life is informed by several analyses of emotions in contentious politics and polarization online (Han 2018; Meng 2018; Sun et al. 2020; Tong 2015; Yang 2009, 2018) and joins a growing body of research that theorizes the role of affect in user practices, including on livestreaming platforms, on dating apps, and in Weibo-enabled protests during COVID-19 (Chan 2021; Tan et al. 2020; Zhang 2023). My goal is to build on such work by examining how affect and emotion circulate in everyday, mundane ways in a number of digital realms, with varying outcomes, among diverse marginalized groups. Outside of China, a large body of research has theorized how affect and emotion flow through different mediated spheres (Garde-Hansen and Gorton 2013; Hillis, Paasonen, and Petit 2015; Karatzogianni and Kuntsman 2012) and undergird various broad-based social movements (Papacharissi 2015).[94] Such research illustrates Garde-Hansen and Gorton's (2013, 31) assertion that emotions are "something that can be distributed and exchanged through psychic and physical contact, but also by social and technological means." These analyses, many textual, are concerned with how affective intensities move people to act, connect, transform, or sometimes disengage.[95] Research in China has also analyzed the ethical issues mentioned earlier (government cover-ups, food safety) as well as certain types of behaviors, such as the spreading of online rumors (Wang 2019) or a particular from of cyber vigilantism called the "human flesh search engine" (*renrou sousuo* 人肉搜索) (Gong and Yang 2017). This book builds upon other ethnographic research with an explicit focus on social media and moral/ethical behavior (cf. McDonald 2016).

Rather than taking the digital realm, or a specific platform, as the starting place for my analysis, I offer a long-term, nuanced account of daily life in all its ordinariness yet extraordinariness, because it is mutually constitutive with social media use such that they cannot be separated. Although Chinese social media can still be the site for a "cacophony" of expressions (Hu 2008), I bring together a heteroglossia (Bakhtin 1981) of marginalized voices in the hopes of enriching our understanding of Chinese people and Chinese social media. This articulation of daily practices, aspirations, desires (the Spinozan "not yet"), affective impulses, emotional investments, and ethical choices are what makes up the assemblage of social media and the ordinary.[96]

A Note on Method

This book is the result of research I conducted over several years, primarily from 2013 to mid-2020. Here I briefly highlight the methods in broad strokes, while further details are provided in each chapter. The bulk of the ethnographic fieldwork in China began in 2013 for several months (although in chapter 2 I draw on some fieldwork from 2011), with additional return visits of varying lengths in 2014, 2015, 2017, and

2019. The fieldwork traversed physical locations in China, specifically Beijing and several villages in Shandong province, and social media platforms—QQ, WeChat, Weibo, Bilibili—though I focus a lot on WeChat because that is what all the participants used. When I began the research for this book, I did not imagine that it would extend over such a long period, but life has a way of intervening, and interfering, in our grandest plans. However, the span of the research enabled me to gain insights into continuities and transformations that otherwise would not have been possible.

Just as life should be understood as in a state of flux, or *becoming*, so too should social media, as anyone who was once a MySpace user can attest. In China, social networking sites that were once ubiquitous when I began my fieldwork, like Renren, have been largely abandoned, and others, like WeChat, which was in a nascent stage, are now so engrained in everyday life that it is nearly impossible to imagine existing without them.[97] While this flux creates challenges for the researcher, it is an un-avoidable attribute of what Postill and Pink (2012) call the "messy web" that is social media. Although a particular platform is clearly important for the affordances it contains (Bucher and Helmond 2018), in attending to my participants' practices, and in keeping with my philosophy of technology that I stated earlier and have articulated in detail elsewhere (Wallis 2013a), my goal is to map the articulations of desires, emotions, ethical investments, feelings, practices, discourses, platforms, and the like that form an assemblage, or "messy web" of social media and the ordinary.

In the same way that the research was multi-sited, the methods employed were multiple. I conducted dozens of face-to-face interviews as well as several interviews over WeChat, and did many follow-up interviews. The research is also informed by informal conversations, in person and online, and countless hours of participant observation at sites including homes, places of employment, NGO spaces, and performances, as well as textual analysis of a range of social media content. One of the beauties of social media is that my research could continue via informal interactions with key informants while I was not in China.

A study that takes ordinary affects and ordinary ethics as its theoretical grounding raises questions as to how one actually "captures" these. This research is not based on a notion of affect as being non-conscious and pre-personal, because that would render it unanalyzable (Tyler

2008). Grounding this study in Chinese understandings of the relational nature of the self, the holistic view of the heart/mind/body, and emotions as intersubjective, I thus turn my attention to the various ways that affect and emotion circulate, intensify, and diminish among bodies in relation to each other (Coleman 2008) and to social media. I do this while "documenting existing practices and affective attachments" (Knudsen and Stage 2015, 16), which in turn are constitutive of ethical self-making in a number of realms—quests for personal aesthetics, entrepreneurship bounded by relational norms, emotional and affective labor, and struggles for gender equality. My analyses are always rooted in the context from which they emerged; that is, the participants' lived daily experiences. Thus, many details included in each chapter might seem tangential to social media use if one believes that technology is only defined as the device or platform. However, as with any ethnographic study, the portraits are only partial. Finally, the chapters in this book, though connected theoretically and analytically, are also meant to form self-sufficient units.

Chapter Overview

Chapter 1 focuses on the experiences of a disparate group of young creatives who migrated to Beijing to pursue a creative endeavor, which they envisioned as a pursuit of "personal aesthetics" (*shenmei* 审美). After situating the chapter in China's developmental shift from "made in China to created in China," and briefly discussing the historical connection between aesthetics and ideology, I explore the young creatives' motivations for moving to Beijing, which were connected to social media, how they developed networks to pursue their creative endeavors, and the ethical underpinning of these desires. I then analyze their processes of aesthetic and ethical self-making through three stages—the "apprentice self," "curated self," and "commercial self"—and the various ways in which these were articulated to social media. The chapter also shows how the young creatives were influenced by local and global discourses regarding the meaning and purpose of creativity, originality, and authenticity. Ultimately, their individual construction of a personal aesthetics was intimately linked to their personal ethics, and both provided them a means of navigating societal and familial expectations while trying to fulfill their dreams.

Chapter 2 examines micro-entrepreneurship in rural China, where state and corporate efforts at informatization are founded on an unwavering belief that technology use and the platform economy will develop the countryside. The chapter is based on fieldwork conducted in rural Shandong during two different time periods: the first took place as digital technology was still diffusing in the early to mid-2010s, and the second after e-commerce had been spreading for some time, in 2019. I focus on family businesses started by couples but actually run by women and analyze whether technology use led to shifts in familial gender dynamics and expanded personal and economic opportunities that could be understood as empowering rural women. I show how women used technology in ways they deemed appropriate and how their entrepreneurial practices were guided by norms of *ganqing*, or feelings and emotional attachment, and *renqing*, which encompasses feelings, ethics, and social obligations and resources. These norms affect everyone, yet women often bear more burden to exhibit appropriate *ganqing* and *renqing* in relationships in which they are subordinate, and their entrepreneurial endeavors are often undervalued or constrained. I argue that although women's micro-entrepreneurship is situated within structural constraints and folded into long-established aged and gendered relational patterns, ethical obligations, and notions of intimacy and trust, through engaging with technology and adopting various tactics, both younger and older women did carve out spaces for agency and achieved some degree of empowerment in the family.

Chapter 3 centers "older" (aged 35 and above) domestic workers, all women, in Beijing, highlighting how they used social media to have a voice, exercise agency, build community, and gain a sense of empowerment. Domestic workers are some of the most marginalized of China's migrant workers. They must contend not only with state and popular discourses that portray them as "backward" and of low *suzhi*, but also with severe constraints on their autonomy in the private home. Engaging with Sara Ahmed's (2014) notion of the "sociality of emotions," I show how the women posted social media content that was often pedagogical, uplifting, and positive, and based on mutual sharing and support. Some women used social media to vent grievances, yet these posts, just like the positive posts, embodied the other-centeredness of Chinese relational ethics. I argue that they used social media to construct themselves as ethical subjects and to create affective value for themselves and each other. Although just as in their jobs their social media use entailed affective labor, social media was also empowering in that it enabled them to construct their own representation.

Chapter 4 unpacks how a loosely networked group of young feminists spread across China channeled various feelings—anger, depression, alienation, pain, and joy—into transformative possibilities for themselves and others as they sought to challenge sexism and misogyny in neo/non-liberal China. Drawing on Sara Ahmed's (2010) notion of the feminist killjoy and ideas regarding "networked publics" (boyd 2010; Ito 2008), I argue that these young women are part of a "networked feminist killjoy assemblage," which is an affective assemblage made up of bodies, discourses, emotions (anger but also hope), objects, technologies, and ethical practices. After discussing the experiences that led them to become feminists, I explore how they wrestled with various namings and meanings of feminism, which in turn informed their feminist praxis online. I then analyze different cultural productions posted on social media aimed at teaching, creating solidarity, and breaking down taboos. The chapter shows how diverse feminists deployed a range of tactics, some that are subtle, some that engage with transnational popular culture, and some that intentionally "stir up trouble" to disrupt the status quo. Because many women face negative consequences for claiming a feminist identity, the final section discusses how they also used social media to regroup and renew.

A brief conclusion summarizes the theoretical and analytical contributions of the book. While recognizing the particularities of the different groups featured in this book and the challenges they faced, I argue for the importance of centering a range of voices and valuing them equally. I discuss the findings of the research in the context of the post-COVID milieu to show how focusing on the entanglement of affect, social media, and the ethical constitution of the self could continue to provide deeper insight into not only disparate marginalized individuals' aspirations, struggles, and small triumphs but also social transformations occurring in neo/non-liberal China.

1

Marginalized Young Creatives, Personal Aesthetics, and the Quest for Meaning

One balmy night in September, I went with two Chinese young creatives, Bo and Lily, to see a punk band at a small club located on one of the straight, wide boulevards in the center of old Beijing.[1] The area had not been gentrified like so many other parts of the city, and the street on which the club was located was lined on each side with rows of individual shops interspersed with entryways to *hutong* (胡同), the narrow alleyways with traditional courtyard houses so characteristic of Beijing. After the show, Bo, Lily, and I went across the street to a small clothing shop they ran, the doors to which a friend had just opened in order to attract some late-night customers. The interior was crammed with used clothing, which also flowed out onto the sidewalk in small heaps that were being picked through by a few of the young, artsy rock fans I had just seen at the show. As we stood nearby, I remarked to Lily that all the clothes seemed to have a "vintage" style, to which she nodded. While we continued chatting, she described a local fashion happening that she had attended a few days earlier, chuckling at the contrast between the high-end designs there and the clothes lying on the sidewalk.

I had met Lily, and through her, Bo, a couple of years earlier at a small upscale clothing and jewelry boutique that she managed in a renovated *hutong,* now a tourist attraction, a few kilometers away from the used clothing shop. Lily was an aspiring fashion designer, and Bo hoped to be a tattoo artist. At the time, he was taking a Saturday class and informally studying designs on Weibo accounts dedicated to tattoos. Back then, Lily used to peruse fashion blogs regularly, especially those focused on Korean style. Hanging out in her shop, when no customers were present, she, a woman employee, and I spent time scrolling through their phones to see which fashions caught our eyes. However, for now she was burnt out on social media. "It puts too much pressure on you to connect," she said, while acknowledging that she should start a WeChat site for her entrepreneurial endeavors because visibility was the key to success.[2] Her ultimate goal was not to become a famous designer just for the sake of it; rather, she wanted to make enough money so that she could start a charity to help others. Still, Bo and Lily achieving their dreams seemed far off for many reasons, not least of which was that living in Beijing as "outsiders" was challenging. Because neither owned a house and they both had rural household registration (*hukou* 户口), they could not get a bank loan, so in order to procure the funds needed to open the clothing shop, Lily had borrowed money from a non-bank lender in what could only be described as a usurious rate.[3] On a tight budget, she and a friend lived in a *hutong* in a tiny single-story brick house (*pingfang* 平房), where her room was so small that it could barely even fit a single bed.[4] Bo lived in a public dorm, sharing a room with seven other men he had never met prior to arriving in Beijing. Yet, they felt the sacrifices were worth it because their artistic pursuits were seen by them as a calling and even an ethical impulse.

With their desire for creative expression, Lily and Bo in many ways are representative of the young creatives now found in major cities all over China, and indeed across the globe, who embrace the "freedom" such work supposedly offers and use social media to learn, network, and promote themselves (or not, in Lily's case) (Chow 2019; Duffy 2017; Lin 2023; McRobbie 2016). Starting in the late nineties, policymakers in several western countries began to deploy "creativity" and "innovation" in the service of economic development, and in the ensuing years this strategy became a transnational phenomenon.[5] Quick to recognize the promise of creativity (*chuangzaoli* 创造力) to bring national and individual transformation, since the early 2000s the Chinese government has invested heavily in regions across the country to realize its ambition to be a cultural creator (Keane 2011, 2013). Indeed, during fieldwork trips in Beijing, I often received so many invitations to film festivals, theatrical performances, and art exhibit openings that I couldn't possibly attend them all. I also met numerous students studying art or design, including some while I was in rural China, as well as white-collar workers employed in conventional "cultural industries" (*wenhua chanye* 文化产业) such as television production, web design, and advertising. Compared to most of them, however, Bo and Lily, and an increasingly large number of young people, are somewhat unique. They come from rural areas and are either self-taught and/or have attended non-elite universities (usually not finishing) or vocational schools. They exist in more marginal spaces, often outside of traditional organizations or industries, and their general life experiences are largely invisible.

In this chapter, I focus on a disparate group of marginalized young creatives ("young creatives" hereafter) who, like Bo and Lily, have migrated to Beijing to pursue a creative endeavor—art, design, or fashion, just to name a few—which they often frame as a parallel pursuit of "personal aesthetics" (*shenmei* 审美). What motivates young people like Bo and Lily to move to Beijing to pursue their creative dreams? How do they develop the skills and networks they need to achieve these dreams and make a living? How are these individualized pursuits, arising from their embodied passions, connected to an ethical impulse to better themselves and society, as in Lily's desire to start a charity? I answer these questions by highlighting the affective flows and ethical choices that are constitutive of their modes of aesthetic self-cultivation as these are articulated to social media.[6] With this optic I hope to add to both the large number of studies on Chinese creative industries that take a political economic perspective (Erni and Fung 2013; Keane 2011a, 2013; Montgomery 2010) and a growing body of rich ethnographic research showing the particular manner in which transnational understandings of the value of creativity and innovation are localized in China (Chong 2020; Chow 2019; Chumley 2016; Lin 2023; Lindtner 2020; L. Zhang 2023).[7] This scholarship has furthered our knowledge of how the authoritarian state's embrace of creativity for value extraction and as a technology of governance, which has increased in neo/non-liberal China, aligns with broader and longer-term efforts to develop the nation and improve the quality (*suzhi* 素质) of its citizens.[8] I build on such work by focusing on the subjective experience of those with less social, economic, and/or cultural capital who are nonetheless interpellated by these newer subjectivities, desires, and aspirations (Rofel 2007), and by highlighting the role of affect, emotion, and ordinary ethics—or the daily practices, judgments, and choices (Lambek 2010; 2015)—in these processes.[9] I examine how these young creatives embrace aesthetics as a motivating force, means of voice/self-expression, and end goal to construct a self in ways that are both antagonistic to and in concert with state discourse. In the process of what Foucault (1997) calls "self-forming activity" with a *telos*, they both resist and reaffirm conventional values regarding what is valued.

After providing background on larger structural transformations and a brief discussion of how historically aesthetics (*meixue* 美学) and ideology have been interrelated in China, I show how for these young creatives, social media provided an inspiration and an affective impulse to embark on a journey of personal growth and a quest for meaning. I then analyze processes of aesthetic self-making as these are articulated to social media through highlighting what I call the "apprentice self," "curated self," and "commercial self," which are not mutually exclusive nor linear in their development. I also reveal how personal aesthetics was constitutive of local and global flows of images and discourses, crosscut by transnational relations of power. As the young creatives simultaneously affirmed and contested normative assumptions about creativity, they also produced hierarchies of taste and distinction. I argue that, ultimately, personal aesthetics was mutually constitutive with personal ethics, providing a moral compass for how to pursue one's passions, negotiate societal expectations, fulfill responsibilities to family members, especially parents, and resist "soul-crushing" commercialization while being able to make a living. For comparison, in this chapter I also occasionally draw upon interviews with white-collar workers who had college degrees and were employed in creative industries.[10]

From "Made in China to Created in China"

Back in 2009, Li Wuwei, a policymaker and academic in Shanghai, published a book about how the creative industries were leading to significant economic, cultural, and social changes in China. He argued that the "upgrading" of various forms of capital—infrastructural, technological, human—would in turn upgrade the nation (Keane 2011b). In the book and elsewhere (W. Li 2009), Li asserted that besides these transformations having a domestic impact, they would also influence how China related to the world. Two years later, his book was translated into English with the title, *How Creativity is Changing China*. In between the original publication and the translation in 2010, Shanghai (the city Li had used as a case study in his book) hosted the World Expo, which showcased not only China's economic achievements but also its creative accomplishments.[11] The China pavilion, for example, presented visitors with a carefully crafted narrative of an ancient culture whose aesthetic and spiritual principles were being harnessed to make China a creator of culture and innovation (Wallis and Balsamo 2016; Wang 2013). In a subsequent event in Shanghai, individual citizens of all ages and backgrounds were invited to be "creative producers" participating in the "creative economy" (He 2011), whether or not they had the means or skills.

For decades, China was known as the world's factory, supposedly able to produce but not create or design. However, particularly since joining the World Trade Organization in 2001, the Chinese government, keenly aware of such critiques, and taking a cue from places like the United Kingdom, United States, and Australia, has sought to transform China's path to economic growth and global competitiveness through an emphasis on "culture" rather than manufacturing. This shift has been discursively constructed as from "Made in China to Created in China" (Keane 2011). A counterpoint to this domestic narrative, however, still prominent in the West, is that China is an imitator, copier, and counterfeiter. From fake Apple stores (Yang 2016; Lee 2011), which are mildly amusing due to their fealty to the original, to ramped-up squabbles between the US and China over alleged intellectual property theft (Brewster 2019), there seems to be no lack of fodder for such a perception.[12] Often held up as the ultimate example of the fake is Dafen Village in southern China, where, since the 1990s, hundreds of painters have churned out millions of copies of paintings by western art masters—Van Gogh, Rembrandt, Klimt—for individual buyers and big box retailers such as Walmart (Paetsch 2006; Wong 2013). The "village" became an art industrial complex, where one dealer had hoped to be the "McDonald's of the art world" with art assembly lines and standardized production stages (Paetsch 2006). In neo/non-liberal China, however, the constant need to upgrade and to innovate has been dramatically transforming Dafen (Arnold 2017).

These conflicting macro narratives of China and creativity, or its lack thereof, seem to be diametrically opposed and to exist in entirely different worlds. However, they both are situated in the government's intensive efforts at cultivating "creativity" and innovation ultimately to accumulate human and economic capital to propel China's rise. They also gesture to the way that culture and the state have always been intricately connected in China and how this symbiotic relationship has morphed because of marketization and privatization. In the mid-2010s, the terms "innovation" (*chuangxin* 创新) and "innovative nation" (*chuangxinxing guojia* 创新型国家) appeared and over time became more pronounced in white papers, development plans, and policy decrees (Keane 2013).[13] Accordingly, the state invested in both traditional "cultural industries," including "main melody" film and television (those that adhere to state ideology), performing arts, and museums, as well as "creative industries" (*chuangyi chanye* 创意产业), such as gaming, animation, and online media (Keane 2011a, 2013; Xiang and Walker 2013).[14] Local governments have designated "cultural clusters" in major cities, with areas devoted to particular industries, such as film or online games as well as art districts, maker spaces, and design houses (Fung and Erni 2013; Lindtner 2020; Keane 2013).[15] Perhaps the culmination of such policies was former Premier Li Keqiang's 2015 proclamation on "mass innovation," "mass making," and "mass entrepreneurship," which encouraged everyone to make and create and to "upgrade" their capacities, ultimately for the sake of the nation (Lindtner 2020; State Council 2015b, 2015c, 2016).[16] As noted in the book's introduction, although these phrases are no longer the buzzwords they once were, their residue is manifest in China's continued investment in its cultural and creative industries and domestic innovation, as well as the state's prioritizing of developing its soft power.[17] Of course, this doesn't mean that everything produced as a result of such policies is actually creative and innovative.

In scholarship focused on western contexts, critiques of the shift from an industrial economy to a creative and service-based economy, and the state's role in these transformations, focus on the concomitant rise of precarity in a number of fields, not least of which is creative work, where discourses such as "Do What You Love" (DWYL) mask how such jobs lack security, are often exploitative, and operate through rigid class, gender, and racial hierarchies (Conor, Gill, and Taylor 2015; Duffy 2017; Gill and Pratt 2008; McRobbie 2016; Morgan and Nelligan 2015). These transformations have resulted from the demise of the Fordist production system, workers unions, and the welfare state, and the rise in their place of neoliberal capitalism, the "gig economy," and the entrepreneurial self (Hardt and Negri 2000; Gill and Platt 2008; McRobbie 2016). The contemporary class of worker in this "social factory" is often called the "precariat" to highlight job conditions characterized by instability, contingency, and exploitation (Gill and Platt 2008) or, in Hardt and Negri's (2000) formulation, the "multitude" to emphasize their potential to be part of new forms of political struggle.[18]

Two points need to be made regarding these analyses and whether they can be mapped onto China. First, as I discussed in the book's introduction, the economic and societal transformations China has undergone, particularly its closure of state-owned enterprises and curtailing of the social safety net after its reform and opening, have certainly produced some similar outcomes. However, these shifts have emerged within a very different social, political, and historical context, which includes China's socialist legacy, rapid industrialization beginning in the 1980s, and urban/rural disparity. Of particular relevance is that, due to China's household registration system, most rural residents have never known the job security and benefits granted to urban dwellers in previous decades—and still provided even now for those who are employed in state-owned enterprises.[19] The young creatives featured here are thus not the victims of a retracted social safety net quite simply because they never had one, and they have more opportunities and greater freedom to seek out work that is interesting and/or meaningful than prior generations of rural residents did. This statement is not meant to downplay the hardships and in some cases exploitation that they face, but it directs attention to their own experiences and reasons for trying to engage in creative work.[20] A second salient point is that the firm link between state ideology and creative output, such as the emphasis on adhering to "main melody" themes and the strict regulation of content, has a long history in China, even if the desired outcome in the contemporary neo/nonliberal milieu is distinct.

Culture and Ideology

An in-depth discussion of traditional Chinese understandings of creativity and aesthetics is far beyond the scope of this chapter, but here I provide a brief overview of both concepts and their interrelationship in order to frame my analysis. Historically, understandings of creativity and aesthetics have been rooted in Chinese philosophy, particularly

Confucianism and Daoism. Confucian teaching posits that aesthetics, or *meixue*, literally the study of beauty, is concerned with goodness, proper moral behavior (*li* 礼), and harmony between humans as well as between humans and nature (Z. Li 2009; Keane 2013). Tu (1985) argues that a more expansive conception of aesthetics connects it to processes of self-cultivation and thus to subjectivity.[21] For centuries, particular modes of self-cultivation were the reserve of a mostly male elite or esoteric few—Confucian literati or Buddhist hermits—even if certain principles and practices diffused among the broader populace (Barry 2003; Sundararajan 2015).

Since the tumultuous early decades of the twentieth century, Chinese aesthetics has gone through various permutations according to shifting political winds. During the New Culture Movement (1910s and 1920s), for example, the content of much art and literature urgently called out injustice and perceived moral decay in order to suggest an ethical path forward.[22] After the founding of the People's Republic in 1949, traditional concepts of aesthetics were tossed aside, along with modern, "bourgeois" notions of "art for art's sake," as creative expression had to adhere to Mao Zedong's dictum that art should portray the standpoint of the "masses of people."[23] Thus followed three decades of revolutionary art and socialist realism for the purposes of ideological control. Early in the reform and opening period in the eighties, a "high culture fever" resulted in intense debates among artists and intellectuals regarding the purpose of art (Wang 1996). Avant-garde artists, while negotiating the limits placed on them by the authoritarian state, still managed to upset traditional notions of aesthetics by producing work that at times evoked violence and discord; in other words, the antithesis of beauty and harmony (Köppel-Yang 2003). After the Tiananmen Square massacre in June 1989, once again the arts were more tightly reined in by the state. This control loosened at times during the Hu Jintao-Wen Jiabao (2002–2012) era, although there were intermittent crackdowns, including the detention of artist-provocateur Ai Weiwei in 2011, and campaigns targeting online content containing "vulgarity," an intentionally vague catch-all term.

As the Chinese government turned toward creativity as part of economic transformation, literary scholar and cultural critic Wang Jing (2004, 13) noted that while creativity might be the "least problematic" issue in the rise of creative industries in western contexts, it was the "thorniest question" in a nation "where creative imagination and content are subjugated to active state surveillance."[24] Wang's words could not be truer in the neo/non-liberal moment, where state control and suppression of alternative ideas has increased even though there is more diversity of expression in individual life. In 2014, Xi Jinping gave a speech that intentionally harkened back to Mao, as he cautioned against "immoral" art and admonished cultural producers to "serve the people and socialism." While targeting "vulgarity," he went further in linking cultural production to his Chinese Dream and to Chinese values, culture, and people's "aesthetic tastes" (Xi 2015).[25] In subsequent years, regulations on cultural content, producers, and celebrities have only gotten stricter and more repressive. If Mao's intent for art was distinctly dogmatic, Xi has kept this ideological thrust while banking on the economic outcomes of China as a cultural producer (Lin 2023).

In tracing this ebb and flow, it is important to point out that, although traditional aesthetic concepts might be debated by elite intellectuals and taught in art school or university literature departments, most ordinary Chinese people likely only have a basic understanding of aesthetics and its connection to beauty and harmony. Moreover, even among those in various realms of art (painting, music, literature), traditional aesthetics has been influenced by several factors, namely globalization, commercialization, and of course the authoritarian state.

Boredom, Passion, and Movement

Although *meixue* is the common term for aesthetics, only a few of the young creatives featured in this chapter had any formal artistic training. In describing their endeavors, they were more likely to use *shenmei*, as mentioned earlier, or *shenmeiguan*. These terms connote less a formal notion of aesthetics than an aesthetic standpoint, taste, or pursuit, which is why I translate *shenmei* as "personal aesthetics." This personal aesthetics is in some ways akin to the self-styling of art students and their involvement in aesthetic communities noted by Chumley (2016). However, it is also different, not only because most of the young creatives lacked such formal training, but also because to them, aesthetics was more of a feeling or sensibility that came from within—an affective force and a "stirring and being stirred" (*gan* 感) (Sundararajan 2014, 183) that motivated their self-expression, interactions with others, and use of social media. Moreover, for many, their pursuit of aesthetics was constitutive of an ethical construction of the self, even if this ethics had varying definitions and manifestations.

The young creatives in general recognized that the state's focus on creativity and innovation for national transformation had created new opportunities and greater acceptance of certain careers. However, most also insisted that they were born with their creative passions, and that feelings and desires, not state discourses and policies, compelled their pursuit of personal aesthetics. To take them at their word means understanding their subjective experience as situated in, but not determined by, the context of neo/non-liberal China. As such, their journeys exemplify Kathleen Stewart's (2007) assertion that affects are not forms of signification but rather forces of circulation that pull a subject in intended and unintended places. I first illustrate this entanglement of passions, possibilities, and movement through the experiences of two informants, Hu Fang and Xiao Sun, separated by space and time, and positioned differently due to not only their gender, but also the opportunities and technology available to them.

A couple of weeks after the punk show, Kai, a young painter, introduced me to Hu Fang, who curated a gallery in Songzhuang, one of the government-sanctioned art districts in Beijing. At 36, Hu Fang was an informal mentor to Kai, and he was the oldest creative I interviewed. As a teen in the mid-nineties, he left his rural village after one year of senior middle school and went to Guangzhou to work in the "world's factory." He was propelled by a desire to "experience the world," he said, and made his way there through kinship networks.[26] This discourse, the networks that took him south, and the poorly paid, exhausting, and sometimes dangerous labor he performed were common experiences of rural migrants during the eighties and nineties and still are even now. Recalling this period, Hu Fang said, "That is where I realized how hard life is."[27] When his relatives started a factory back home, he returned there to work yet was soon dissatisfied. Encouraged by an uncle who recognized his artistic skill and ambition, Hu Fang began studying one-on-one with an artist. "I really wanted to become a painter, so I moved to Beijing," he told me. There he lived on the outskirts of the city in an area that had become a haven for artists who had been displaced after they

Figure 1.1. A mural at 798 Art District in Beijing. Photograph by the author.

were forced to leave the makeshift art colony they had created at the Old Summer Palace (*Yuanmingyuan* 圆明园).[28] Little did Hu Fang know this locale would eventually be the home of 798, Beijing's first "official" art district, which later became a booming commercial area (see figure 1.1). Over the next several years, Hu Fang moved between various creative jobs—magazine layout, web design, CD artwork for rock bands—until he finally landed at a "state-owned cultural enterprise" (SOCE) in 2008. Eventually, in 2013, he and like-minded artistic collaborators opened a gallery space in Songzhuang.[29]

The same year Hu Fang founded his gallery, I met a young woman named Xiao Sun, who was working in a boutique a few doors down from Lily's shop. She had grown up in a village a couple of hours from Shanghai and had become a kindergarten teacher in a small town nearby, but her real passion was dancing. As a teenager, she had been an avid user of QQ (a chat application) and Qzone (a mobile social networking app). In 2011, when she was 19, Xiao Sun came across a dance teacher's videos on Qzone. Talking about the videos, she excitedly said, "I was so *moved*! You can't fake that! They were so good!"[30] This affective force (*gan*) literally led to her desire to move to Beijing, where the teacher had a school, so that she could pursue her dream of being a dancer. Xiao Sun saved some money, and a year later her brother rode with her on the train to Beijing, helped her check in to a cheap hotel, and after assisting with the tuition fees for the school, left the next day.[31]

Hu Fang's and Xiao Sun's narratives certainly share some commonalities, and in many ways are quite ordinary. They both had a rural upbringing, although Hu Fang's was more impoverished. They also expressed to me how they saw themselves as always different from their rural peers growing up, in terms of their personalities, values, and desires. Moreover, their movements were also enabled by China's development patterns—from building the "world's factory" to harnessing creative labor for economic transformation. Still, Hu Fang's gender clearly played a role in his greater mobility and ability to build the social networks that led to his position as a curator. Another big difference was in how their embodied passions were given flight. Hu Fang did not have internet access as a teen in his village, but he knew "there was something more." Right around the time he left, the government, in line with its economic imperatives, was ardently trying to diffuse the internet to rural areas. It was simultaneously expanding various types of higher education and revising the educational curriculum at all levels to include more room for creative thinking (Chumley 2016; Kipnis 2011). By the time Xiao Sun was a teenager, China's technological advancements had had a profound influence on her life, even if the same cannot be said of its educational reforms.

Xiao Sun's narrative, and the role of social media in it, is representative of how nearly all the young creatives in this chapter felt compelled to move to Beijing. Many had started with a somewhat uninspiring educational or career path, felt stifled, and were motivated to pursue a creative field through something they viewed on a social media platform, exemplifying how visual media, articulated to personal desires and contexts, can evoke a powerful felt quality that not only stirs people emotionally, but also literally propels movement. Wilson, an aspiring filmmaker from a small rural town in Shandong province, was in a WeChat group with fellow artists. After one of them shared a post from an NGO about a documentary video project, he went to Beijing to participate. In Bo's case, early on it was Douban, a social networking site focused on film, literature, art, and music. "Before Douban," he said, "I felt it was hard to meet people like me, so

I was inspired."[32] After discovering tattoo design through various Weibo accounts, Bo moved to Beijing to become a tattoo artist. Unlike Hu Fang and earlier labor migrants, who invariably relied on kinship networks to facilitate their migration, Xiao Sun and other young creatives I knew often arrived in Beijing with no job, little money, and no, or in some cases, just one or two contacts.

Linking the young creatives' movement to social media is not meant to invoke technological determinism, nor is it meant to downplay the amount of social media content that is problematic at best. Rather, my goal is to highlight their understanding of their subjective feelings, as these were articulated to social media, in propelling action. Their embodied experience was also expressed in their refusal to believe that their passions were the result of a technique of governance, in which the current iteration of long-term discourses regarding the intertwining of the quality of the people and the quality of the nation manifests as the capture of creativity and innovation to fuel China's rise and the realization of the Chinese Dream. Bo, Xiao Sun, Wilson, and others experienced pleasure, hope, and excitement for the future through engaging with social media, which was inseparable from feelings of disdain for and boredom with their current circumstances.[33] They and others like them were supposed to be the beneficiaries of the educational reforms mentioned above, yet those who enrolled in some type of formal course were often soon disillusioned, feeling that what they were taught was impractical, uninspiring, or otherwise subpar. Lily had studied for two years at what she called a run-of-the-mill university but then dropped out. "It was very impractical and kind of boring," she said.[34] She had been following Weibo accounts dedicated to fashion (both Chinese and foreign) and decided to move to Beijing to pursue a dream of making it as a designer in the industry. Once in Beijing, she was drawn to one of the smaller, commercial *hutong* areas because of what she found there—local designer fashions, handmade crafts, and stylish tea shops. Bo, too, had dropped out. In Beijing he had considered going back to school to further study graphic design but decided not to, saying:

> There are too many people who know Photoshop, so you don't make a lot of money doing that kind of work and you don't have a very high social position. A middle school graduate can make between 2000 and 3000

yuan per month in a print shop. The employees are mostly migrant workers who haven't studied, and even if you've studied the outcome might be
the same. It makes me feel distressed.[35]

Bo referred to migrant workers with a middle school education,
who do make up a large segment of low-paid service workers, yet there
are also hundreds of thousands of young people from China's second
and third tier cities and rural areas who, after the educational reforms,
pursued a degree, seeing it as a path to betterment. The result was the
so-called "ant tribe" (*yizu* 蚁族), a term used for college graduates who
migrate to large urban areas to try to make a life and place for themselves. Many do not succeed and end up living on the outskirts of cities like Beijing, unemployed or underemployed and impoverished (Kan
2013; Suda 2016); hence, the reason for Bo's "distress."[36] Once when this
topic came up, Kai, with a clear tone of disdain in his voice, said:

> Those college graduates come here all having the same skills and wanting the
> same thing. There is nothing to differentiate them. Also, they don't realize
> that they probably aren't that skilled compared to others. They want a certain
> thing and don't seek other opportunities or realize their own limitations.[37]

Kai's judgment seems harsh, yet in further conversations, Bo, Lily, and
other young creatives leveled the same critique often heard in the West
against the Chinese education system—there is too much rote learning,
little independent thinking, and no creativity.[38] To them, many of those
with university degrees and employed in creative industries had merely
learned some skills but did not have real knowledge. Even worse, they
were opportunistic, taking a seemingly easy path to secure a decent job,
though with "no passion" and no concept of personal aesthetics.[39] The
young creatives *did* care about earning a living and their social position,
yet just as they differentiated themselves by their motivations, they also
sought alternative means to reach their goals.

Personal Aesthetics, Social Media, and the Cultivated Self

Having left behind an unfulfilling, "boring" education or job and
embarked on a journey to pursue their passions, the young creatives were

deeply invested in cultivating knowledge they felt was meaningful. Here I discuss how social media did not just provide inspiration but was inseparable from their cultivation of their personal *shenmei*, or their working on the self with a desired end (Foucault 1990, 1997).[40] Such work on the self was manifest in their development of the apprentice self, curated self, and commercial self, which, as mentioned earlier, were not mutually distinct. Rather, these were intertwined processes of aesthetic self-making that I separate for analytical purposes.

The Apprentice Self

As apprentices, the young creatives were devoted to gaining knowledge and skills through formal and informal means. In the process, they embraced some of the modes of learning they so ardently critiqued as part of the problem with China's formal educational system, while simultaneously rejecting outside accusations that China represents the antithesis of creativity. I first situate these practices within historical understandings of learning and creativity in China.

In the Chinese tradition of education, value is placed on emulating role models. This mode of learning survived during the Mao era, with revolutionary heroes replacing those found in ancient legends, and it continues today in formal education, state discourse, and popular entertainment, with Confucian sages, patriotic martyrs (like Lei Feng, a model soldier from the 1960s who just won't seem to fade away [see figure 1.2]), celebrities, and other public figures held up as exemplars.[41] Although the idea of humbly learning from one's masters exists in diverse cultures, in China the party-state's selection of widely propagated role models is often seen as a means of social control. Thus, for years those who have sought alternative viewpoints have turned to the online realm. Even as state control has intensified, social media is still a space for divergent views, often shared through coded language, humorous memes, and other techniques (Yang 2009; Zeng 2020).

The penchant for emulation in learning is often mentioned in the critiques of China as a copycat nation mentioned earlier. What is usually missing from such critiques is that understandings of creativity in China have their own roots and are based in a different epistemology than that of the West. That is, historically and from distinct philosophical

Figure 1.2. A poster urges passersby to carry forward Lei Feng's spirit. Photograph by the author.

traditions, creativity has been viewed not as a singular accomplishment but, in keeping with the relational understanding of the self, as co-constructed (Ames and Hall 2003; Keane 2019). Sundararajan (2015) adds that Chinese creativity is "intrinsic-self-oriented," meaning that self-growth as part of the creative process is important. This idea is true in western conceptualizations as well; however, she argues that the Chinese "self-reflexivity principle" causes the ultimate goal to be different. Unlike in the West, where "patents or other measures of product" dominate, historically "the measure of creativity in China has consistently been self-transformation of the creative individual him- or herself" (145). The party-state's latching on to creativity and innovation to advance China economically and politically is an instrumental goal and does not negate this fundamental notion of creativity as it relates to the inner realm of a creator.

"SOCIAL MEDIA CAN COMMUNICATE GOOD THINGS"

As apprentices, the young creatives engaged with historical and cultural understandings of learning, and their pursuit of personal aesthetics seemed rooted in the Chinese philosophical traditions mentioned above, even if not always consciously expressed in that manner. Most had found informal mentors, who took them under their wings: Hu Fang for Kai, a Canadian expat boss for Lily, a woman gallery owner for Xiao Sun, and an expat friend for Kyle, who did interior design. These mentors played an important role in fostering their growth and self-cultivation. However, because mentors only have so much time to give, the young creatives also learned through alternative means, namely social media platforms. They didn't ignore the negative aspects of social media, but as one young woman who was a web designer remarked, "Social media can communicate good things; [it] can raise people's aesthetic level (*shenmei shuiping* 审美水平). It's not like [Chinese] television," which is heavily regulated by the state.[42]

Social media platforms, of course, provide educational content, formally and informally, on everything imaginable. Across the chapters in this book, various participants engaged in some manner of informal learning online, and in their focus on gaining practical knowledge, the young creatives were similar. They were avid viewers of online lecture series, followers of influential figures on Weibo, and readers of public

WeChat accounts devoted to various creative endeavors—dance, music, fashion, and drawing.[43] For example, Kyle, who had completed some vocational training, learned much about how to renovate courtyard houses partly by religiously watching the BBC show *Restoration Home* through a video app. It might seem ironic that he would find a British reality television show helpful for restoring some of the most historic dwellings in China. However, Kyle explained, "I studied the episodes, and I learned a lot about the process [of historical renovation] in general. You can gain skills. It doesn't matter where [the actual building] is located."[44] Bo, after deciding not to go back to formal education to study graphic design, took a tattoo drawing class on Saturdays. As we discussed his class and his hoped-for future career, he said he enjoyed the class but believed he also learned nearly as much through utilizing online resources. Besides practical knowledge, the young creatives also sought out ideas that "fed their soul," such as Chinese philosophy and information on contemporary social issues.

COPYING AND CREATION

As the young creatives developed their apprentice self, they were clearly aware of how China, and by extension Chinese people, were cast as unimaginative and lacking creativity. Their rejection of these discourses says as much about Anglocentric notions of originality as it does about actual practices, as I illustrate through the experiences of Bo and another tattoo artist apprentice. During the same conversation in which Bo discussed his tattoo learning in class and via social media, he said, "I look at different Weibo accounts and different designs, both Chinese and foreign [Korean, Japanese, and American]. I copy some of these into my sketchbook." He added, "I then try to come up with my own designs."[45] While talking, he scrolled through Weibo, showing me the various accounts he followed (or those of which he was a *fensi* 粉丝, or "fan"). He then flipped through his sketchbook, which contained an impressive array of drawings of everything from motorcycles to what appeared to be mythical creatures. Yuki, who was learning to draw tattoos from her boyfriend, also relied on social media, including Weibo, Baidu Tieba (an online forum), and Instagram. In one conversation, she said, "I follow Italian, Korean, Japanese, and Spanish (individual) accounts on Instagram. I like them better. Chinese tattoo artists copy them."[46] She also discussed how she and other tattoo artists she knew shared designs, both their own and others, with each other.

In conversations with Bo and Yuki, at different times I raised the critique of China as copycat nation. Bo smiled but also seemed a little bothered and replied, "Everyone copies," and then added that nothing is truly original. Regarding tattoos in particular, he saw some irony in the fact that western tattoo artists are engaging in an art form that originated elsewhere and they liberally lift images from China—especially dragons and Chinese characters (often incorrectly)—yet no one accuses them of copying.[47] Yuki, on the other hand, took my question in stride. She said, "Tattooing is described in books like the four great classical novels (*Sida mingzhu*)," implying that other cultures copied Chinese art forms, and she saw no problem at all with copying.[48] Separately, she and Bo expressed the impossibility of disentangling the copy from the original within the transnational circulation of images.[49] On the surface level then, it would be easy to assume that, despite their protestations about the opportunistic art majors, who supposedly have no creativity, in some ways they were no different—yet this conclusion is too simplistic.

Outside China, a large body of scholarly work has engaged with how digital culture troubles the notions of originality, creativity, and copying that Bo, Yuki, and other young creatives also raised. Much of this research examines the commercial, legal, cultural, and political aspects of what is variously called "remix" (Lessing 2008), "configurable" (Sinnreich 2010), "participatory" (Jenkins et al. 2009), or "spreadable" (Jenkins, Ford, and Green 2013) culture. However, because it is located in western contexts, these practices are often positioned in opposition to the Enlightenment (and Romantic) construct of the autonomous, individual genius creator, who is endowed with a singular, unique mindset that enables him (pronoun intentional) to create something truly original.[50] In contrast, Sundararajan (2015) states that because China does not have a creation myth, there is "no need for heroic narratives of revolutionary creativity" or for "causal accounts of creativity" (144).[51] Moreover, historically, creativity has been viewed as based on authenticity (to oneself) rather than uniqueness or novelty (Sundararajan and Averill 2007). The young creatives' practices and ideas about learning, copying, and sharing are clearly rooted in these ideas, again even if they hadn't studied them directly. Sharing was part of their everyday ethical practice, undergirded by their passions and desire for growth.

Ultimately, ideas about creativity and practices around learning, sharing, and copying had an endpoint: crafting a personal aesthetics. Social media platforms of all kinds were used as a means to an end in this regard. Moreover, the expression of the young creatives' personal aesthetics manifested in how they curated themselves online, which was governed by ideas about taste, standards, and distinction, as well as norms of relational ethics, as I discuss next.

The Curated Self

In some sense, everyone who uses social media curates a particular image of themselves, as an abundance of scholarly literature has shown (Chua and Chang 2016; Senft and Baym 2015) and as I demonstrate across the chapters in this book. In the realm of US creative work, Duffy (2017) has revealed how content creators in the fashion industry struggle to perform an outward identity that balances "authenticity" and "self-promotion." However, the young creatives featured here were not trying to gain massive followings and become influencers, or what are sometimes in China called online "key opinion leaders" (KOLs), although they did struggle with issues of self-promotion, as I discuss later. Here I show how their curated self was an extension of their apprentice self, revealing another facet of their continuous processes of self-making as a manifestation of their embodied passions and quest for personal aesthetics. As such, although they did not necessarily know each other, they shared many similar norms and beliefs that influenced their ethics of self-presentation online.

ON WHAT NOT TO POST

Much scholarship has been devoted to showing how style, taste, and aesthetics are unifying signifiers for a group as well as a means of distinguishing oneself, often through negation of others' taste, style, etc. (Bourdieu 1984; Thornton 1996). The cultivation of a particular aesthetic sensibility informed how the young creatives curated themselves online, with distinction gained at times in this manner (e.g., differentiation and negation). For example, Bo contrasted himself with his rural peers who had graduated from university, returned home, and were playing it safe. He said:

> They are married and have a kid or two, their parents gave them a house, they have a job that isn't demanding and when they are not working, they waste all their time playing video games. They don't seek challenges in their life. I have nothing in common with them. Their posts on Weibo or WeChat are just about the food they cooked and/or ate, their family, or cute animals. There's nothing that stretches them or causes them to think critically or learn.[52]

Bo then quickly scrolled through his WeChat Moments, showing me some of their posts, and it was hard to disagree with his description—there were indeed many images of food, children, and pets—although I didn't necessarily share his assessment. Bo might have been the most vocal in his critique, but Xiao Sun, Lily, Yuki, and most of the other young creatives voiced similar sentiments in different ways. Xiao Lin said his classmates back home posted "stupid stuff. It's boring. Their emotional intelligence is low."[53]

Aside from revealing taste by negation, these critiques of "boring" or "stupid" posts are another manifestation of the ubiquitous refrain that so many things in rural China are dull and meaningless, a feeling that propelled the young creatives' movement in the first place. That these same affective forces anchored rural peers to their villages, providing safety and security, "proved" to Bo and others that they not only lacked ambition but also, more importantly, passion. Still, the young creatives' equating of the countryside with stasis mimics government and popular discourse, and in one sense can be understood as a manifestation of how such discourse has effectively colonized consciousness (Yan 2008) and harnessed affect as a technology of governance (e.g., activating individuals to be creative and/or entrepreneurial to escape boredom and stasis). The young creatives were clearly influenced by these discourses and forces. At the same time, they faced very real structural impediments back home, where there was little space for creative expression.

MODESTY AND ANONYMITY

Given the critiques of the mundane things that should *not* be posted, it is logical to assume that Weibo, WeChat, and other social media platforms would be prime venues for the young creatives to display their talents and/or engage in self-promotion, yet this was not exactly the case. In the process of learning, studying, and crafting their personal aesthetics, many of the young creatives seemed self-conscious of their own abilities and limitations and were quite careful in deciding what or if to post and where to display their creative endeavors. Violation of these protocols sometimes invoked extreme judgment. For this reason, social media platforms that offered a degree of anonymity were taken advantage of by some to post their art or designs. For example, for Yuki and several others, Weibo was "freer," meaning it was a platform where, using a pseudonym, one could post things (hopefully) free of judgment from those whose judgment mattered most. However, while Yuki wrote short diary-like entries on Weibo, she did not post tattoos she had drawn there, at least not when we first met.[54] Her reasoning was simple: "I don't want others to know me that deeply."[55] She posted on Instagram instead, because it was blocked in China and thus less likely to be viewed by many people. As she showed me some of her tattoo designs, both on her iPad and on Instagram, she said modestly, "They are not that good," even though some had received several likes. Such self-effacement, demonstrating modesty, was common. Similar to Yuki, early on Bo used his phone's photo album (rather than social media) to archive a mix of his own drawings with drawings from which he sought inspiration.

A combination of modesty and anonymity also informed the opposite practice—taking advantage of social media potentially to gain informal critiques, yet with varying degrees of success. One artist from Sichuan posted some drawings on Weibo, but she was disappointed when they did not garner much attention. Cui, another young artist, stated, "Since no one knows it's me, it's okay. I can put stuff out there and see if I get comments, likes, or fans."[56] For Cui, anonymity was a key element not only for gathering feedback, but also for avoiding potential embarrassment if friends, colleagues, or competitors did not like what he posted or were to see someone else's critique. When asked if he posted his designs on WeChat, where he knew everyone in his contact list, his answer was "never." He explained:

On Weibo and WeChat you can learn a lot of things and be stretched and challenged. Then you compare your level of ability to others and realize you are not as good and need to keep growing. If you don't have delusions about your ability, this is how it is. You realize there are others a lot higher than you [in skill level], so you don't want to show off foolishly.[57]

In stating that one should not "show off foolishly," Cui was alluding to the importance of face (*mianzi* 面子), namely not losing face by posting something that did not demonstrate enough skill and then being subject to a critique. Most agreed that those who did post their designs on a more personal platform like WeChat before reaching what others felt was a certain level of accomplishment risked being looked down upon. For these reasons, Xiao Sun, the dancer, shared different aspects of her life on different social media accounts. Initially she had had two QQ accounts, which eventually morphed into two WeChat accounts. In the WeChat group for close friends and family, she posted videos of her dancing so that only they could watch. In these various practices, privacy, anonymity, and modesty were bound together.

SUBTLETY AND SUGGESTIVENESS

If anonymous posts were one way to express humility in curating the self, posts on WeChat, as Cui mentioned, could potentially be viewed by all WeChat contacts.[58] A curated self that was visible, however, did not mean the young creatives filled their WeChat Moments or other social media with their artistic creations or performances. Those who did that were perceived as being too full of themselves and, as discussed in the book's introduction, as ignoring interpersonal ethics by possibly making others feel bad or embarrassed if they did not have a similar level of skill, which relates to the norms of modesty just discussed. For many, a subtle approach was preferred, consciously or not evoking something similar to what has been called aesthetic "suggestiveness" (*anshi* 暗示), or "to show by indirectly hinting at it," a prominent aesthetic property of Chinese poetry, literature, and painting (Gu 2003, 490).[59]

This suggestiveness manifested in a number of ways, one of which was to post images with minimal or even zero context provided. For example, many of Lily's posts on WeChat presented daily life—pictures she took of an interesting view of the night sky, flowers wrapped in paper, a glistening walkway after a rain—with no description. Under one image of her *hutong* courtyard with two plants in pots and part of a wicker chair in the frame, she wrote "remembrance" (*jinian* 纪念). After Xiao Sun stopped dancing and threw herself into the art world (discussed in the next section), she only used one WeChat account, which took on a much more carefully curated form. Like Lily, she also posted pictures of daily life, such as her (presumably) walking on a crowded sidewalk during rush hour with the comment "morning." Other posts included a picture of her desk and the ceiling at the gallery where she worked (I only knew where these photos were taken because I had visited her there).[60] Regarding such posts, Xiao Sun said she wanted to express a "feeling."

A second way such suggestiveness manifested was by providing a running archive of the art and artists, musicians, dancers, etc. that inspired oneself, again often, but not always, with minimal comments depending on the topic. For the first two years I knew Bo, his Moments were infrequent and were mostly of songs of his favorite bands, a few images or videos of live music shows he attended, write-ups of artists he liked, and sometimes pictures he took on the street (he loved photography and on one occasion gave me a printed booklet of photos he had taken). Wilson, the filmmaker, posted a mixture of articles on movie directors and writers, yet he tended to reserve comments and discussion of these for a private WeChat group with other artists. Unfortunately, over time this group was shuttered for having "sensitive" content and had to be recreated several times.[61] Kai was the only young creative who said he used WeChat solely for communicating with friends and subscribing to several public accounts. "We don't need more pictures on WeChat!" he had insisted.[62] When I first knew him, he said he liked to fill his Moments with forwarded articles on art and artists that he found inspiring.

A third manner of suggestiveness was through interspersing posts of friends' creative endeavors with one's own so that the latter was relatively inconspicuous. For example, Lily made posts that promoted local designers, such as an image of her wearing a blouse, designed by her "friend, mentor, and hero," and made of "her hand-made cloth

[that] makes me so warm in this cold weather." Another example was an amusing series of eight pictures with a different piece of handmade jewelry, such as a ring or earing, displayed on a cat (who appeared unfazed). Without having visited Lily's store, it would be impossible to know which fashions or jewelry she had created. After Bo began working in a design firm in 2018, gradually his Moments started to include occasional drawings of his own, yet without captions, leaving it to viewers to surmise if he was the artist. Kai also went through such a transformation after he got a job in a gallery (with the help of Hu Fang). He started posting pictures of paintings and photos he had taken, interspersed with those of other artists. Again, it wasn't always possible to know for certain which were his own. All of these practices combined modesty and subtlety to inform, and perform, one's personal aesthetics.

AN ETHIC OF SOCIAL CRITIQUE

One day Kai posted on WeChat a short narrative that accompanied pictures he had taken after going back to his rural village and seeing that his elementary school had been demolished to make room for new development. Rather than focusing on the destruction, his photos showed a wall glistening in the sun and trees with birds perched on the branches. Below the photos he included a moving remembrance of his childhood, which was one of the few times he wrote something so lengthy. This post is representative of a final distinctive aspect of the young creatives' curated self, which was social critique, such as calling out inequality and the need for social justice. Like Kai's remembrance, such posts demonstrated an ethical imperative, in some cases more explicit than others.

The young creatives tended to say they were interested in "social" rather than "political" issues although sometimes there was a fine line between the two.[63] These issues tended to be ones with which they had some sort of personal experience or connection. For example, Kai posted his mixed feelings about rural development, and Lily forwarded articles on left-behind children in rural villages and rural poverty.[64] Wilson posted about the notorious "urban management officers" (*chengguan* 城管) abusing their power, which he experienced firsthand when he made a film about night street vendors in Beijing.[65] Other young creatives posted indirect critiques. One article shared with me was on

Milan Kundera, but it was later blocked, apparently for being critical of communist regimes.

In contrast, although some of the white-collar young creatives said they posted occasionally about social issues, it seemed that those who were working corporate jobs treated their social media, especially WeChat, quite differently than the marginalized young creatives. For example, Kit, who was employed by a foreign advertising company posted frequently, and he showed me several pictures of mixed drinks he enjoyed after work and foreign sports cars along with sports news. Susan, who worked for a fashion company, had many posts of content from western and Chinese fashion magazines (*Vogue, Marie Claire*), celebrities (Brad Pitt, Zhang Liang), and about being overworked—what came to be known as the infamous 996 work schedule (meaning one is expected to work six days a week from 9:00am to 9:00pm).[66] Although my interactions with them were limited, these creatives seemed much more like the young people employed in SOCEs that Anthony Fung (2016) called "contented bourgeoisie": talented, constrained by corporate culture, not as imaginative, and seemingly content with a "settled" life. Perhaps Dan, who was an editor for one of China Central Television's (CCTV) channels, epitomized this attitude most, when during our interview the issue of inequality in China and the potential for social media to hold the government accountable to deal with it came up. "Maybe social media can help change China," he said, but "anyway, I don't want Chinese society to change that much."[67]

The Commercial Self

In neo/non-liberal China, entrepreneurialism, innovation, and individual drive are heralded as necessary to the nation's future and for achieving the Chinese Dream, yet the young creatives had contradictory feelings about creativity and commercialism. Many railed against the invasion and hollowing out of aesthetic life by processes of economization, or in the words of Kai, "Materialism is valued above all else."[68] They acknowledged, however, that some of their creative friends did not necessarily feel this way. They also recognized that these processes had created more acceptance of and spaces for creative work. For example, Kai's gallery job was in one of the vast art districts, which exist because of heavy government investment. The increasing emphasis on profit in such spaces has dramatically transformed them from places for iconoclastic and even antagonistic artistic expression with less overt commercialization to an overpopulated mélange of upscale coffee shops, stores full of kitsch, and galleries selling art at exorbitant prices.[69] Kai asserted he could still make a living and be true to himself, a struggle that has been experienced by artists across time and space. Here I discuss strategies the young creatives used online and off to balance commercial demands with their own values and personal ethics and aesthetics.

SHARING VERSUS ADVERTISING

One tension the young creatives who were earning a living from their creative pursuits faced was how to share something online without it being perceived as self-promotion. In their networks, the latter was seen as very distasteful, especially if done with friends and family.[70] In Lily's case, without having ever visited her boutique, it would be impossible for someone to know from her posts which fashions or jewelry she designed or sold, as mentioned above, and that was her preference. Yuki, who did not want people to know her too "deeply," later expressed another reason for her hesitancy to post her designs on WeChat: "People think you are showing off. I don't want them to think I am advertising."[71] Still, there were a few who used WeChat for this purpose, such as Jade, who designed some of the jewelry in Lily's shop, and Xiao Sun, whom I discuss below. Other apps were deployed as well.[72] For example, Yuki had previously used Momo to find clients, which I found quite surprising because it is primarily known as a hook-up app (like Tinder) even though the company has tried to change its reputation.[73] Yuki said she gained clients this way, but she also received a lot of sexual solicitations and eventually stopped using the app.

While eschewing direct forms of self-promotion, many of the young creatives deemed it appropriate to publicize events where they could earn some money while helping out a charitable cause. Kai had made promotional films, which he posted on WeChat, for a project started by an artist from the United States who brought together other artists to paint different objects (such as a car) that were then auctioned off for charity. Lily had a strong conviction that her interest in fashion design should be used for the greater good. She had previously worked in a clothing shop owned by a Canadian who was very involved in charitable endeavors. At the time, private charity was something quite new and underdeveloped in China. Lily was deeply moved by his heart for impoverished people and hoped she could be successful enough to be able to give to others the way that he did. On her WeChat Moments, she frequently posted about charity events that were organized to mutually benefit designers and those in need, such as "Charity Bring-your-own-clothes Pop up Shop." Another post was about a foreign friend who connected his fashion brand to promoting sustainable agriculture in China. Lily also sometimes organized charity events even though she barely made enough money to survive. It did not appear that any of these efforts were seen as a personal sacrifice or as a contrast with state-directed volunteerism (see Ning and Palmer 2020) but rather a "natural" joining of their personal aesthetics to an ethical self-cultivation that involved helping others.

PERSONAL AESTHETICS VERSUS FAMILIAL EXPECTATIONS

Some young creatives seemed to struggle more than others with balancing the tensions between their moral convictions and earning an income. The first time I met Wilson, he was filming an elderly blind man who was playing the *erhu* on a street corner.[74] As mentioned earlier, Wilson had moved to Beijing to take part in a documentary project organized by an NGO, which had designated four teams that were each supposed to film a different "low-social status" group. Prior to arriving in Beijing, Wilson had saved money by doing several odd jobs—teaching guitar, driving a truck, and interning in a local "creative company," which he quit even though he was offered a full-time job because he felt "they wanted to control too much what I could say." He had also studied Chinese philosophy on his own. I asked him at one point if he was concerned about his lack of resources/job and he said, "The most important thing isn't money, it's to do something meaningful . . . something you feel [have a passion for]."[75] Once he got married and became a father, however, he returned to Shandong province and got a full-time job. Even then, he still found a way to pursue his passion, eventually winning a prize in a film festival for a documentary he made.

Wilson's return home relates to a final aspect of the young creatives' "commercial self," which was the tension between doing work that enabled them to remain true to their ethics and personal aesthetics while fulfilling obligations to family. In neo/non-liberal China, personal happiness, individualism, and the pursuit of self-interest are no longer scorned in the way they were in the past, as discussed in the book's introduction. However, there are still very distinct notions regarding how a child demonstrates they are "filial," namely, by securing a job that gives "face" to the family and that provides the financial security that ensures parents will be supported in their old age. In Yuki's case, becoming a tattoo artist is a relatively new career path in China and one that is often viewed with suspicion. It is also not occupied by many women. Not surprisingly, Yuki's parents were not happy when she told them about her work. Her father in particular associated tattoos with "hooligans," an attitude Yuki labeled "traditional." She said, "[My parents] still want me to find a more stable job. They want me to work in the government or become a teacher, but I dislike those kinds of jobs. I like the freedom that comes with this job."[76]

As discussed earlier, a critique of the discourses of "freedom" and DWYL is that they mobilize people's passions for the sake of corporate as well as state capital accumulation. In China, those with college degrees who are employed in jobs in conventional creative industries tend to have better pay and working conditions (Keane 2016). The young creatives featured in this chapter did not have such financial security when I met them, but they felt pursuing their dreams was worth it, at least while they were young. For example, Leo, a struggling photographer, took advantage of the lack of security in his job by quitting after saving up enough money to bicycle from Beijing to Lhasa. After he made this decision, his parents made him feel a bit guilty for his "selfishness."[77] Like Leo, those who lack a steady income as a result of their creative pursuits often feel pressure from family, and some friends, to justify such "selfishness."[78] The young creatives balanced these tensions through linking their creative endeavors and personal aesthetics to a personal ethics that spoke to a realm beyond individual desire and was rather about making the world a better place. Yuki thought her job was quite meaningful. "A tattoo will follow the person their entire life; they often are part of the person's story," she said.[79] She earnestly sought to design patterns based on what customers wanted, but this too clashed at times with her own values, such as when a client from the United States asked her to draw a panda holding guns (see figure 1.3).

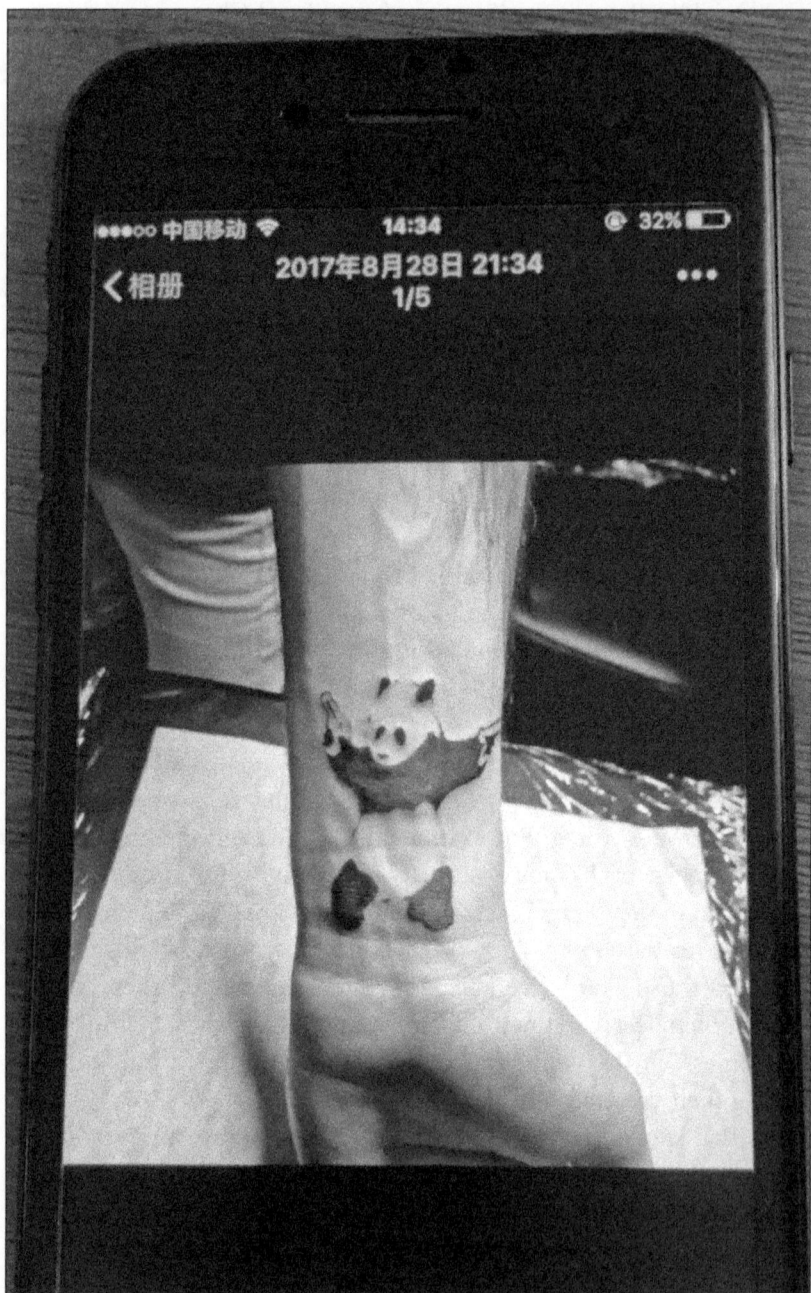

Figure 1.3. When marketplace demands don't align with an artist's personal ethics. Photograph by the author.

This discussion is not to imply, however, that personal ethics and aesthetics have a necessary connection to charity or altruism. Xiao Sun eventually gave up dancing except as a hobby; however, she parlayed the creative connections she had made into eventually working at a gallery in Beijing. A successful businesswoman owned the gallery and had taken Xiao Sun under her wing to mentor her. Over coffee one day as we talked about her new path, Xiao Sun expressed concern that she was getting older and that she had to think about her future—marriage and kids and helping her brother support her parents. During our conversation, she told me her desire was to be like a famous American auto salesperson whom she had read about online. Although she repeated his name and wrote it in Chinese, I had no idea who he was. It was not until I looked up the name that I understood: Joe Girard, a former Chevrolet employee cited in the *Guinness Book of World Records* for selling the most cars in a year, and who later became an author and motivational speaker. Xiao Sun wanted to be an extremely successful art entrepreneur, even if this meant comparing her job to a car salesman, which brings us back full circle to the art dealer in Dafen Village, who wanted to "McDonaldize" production and sales of copied paintings.[80] Here indeed was a different sort of ethical imperative that motivates desire.

After our conversation, Xiao Sun began to focus all of her energy on learning about art, and especially the art trade. After she got the job in the gallery, over time her WeChat posts transformed from suggestiveness to posts about upcoming art exhibitions and photos of receptions at the gallery and artists she had met. Eventually, these were outnumbered by images of just the art, sometimes accompanied by poetic, somewhat flowery language. For example, above six images of an installation of glass sculptures she wrote, "Ice is nature's son, flowers are nature's soul. Ingenious thinking and the combination of chemicals and materials gives rise to the world's most beautiful moment." When I next saw Xiao Sun on a subsequent stay in Beijing, she had moved to a different, more prestigious gallery and had met many artists and helped organize exhibitions. As she acquired a certain type of cultural capital, her taste had shifted accordingly. When I mentioned Joe Girard, she was visibly embarrassed that I had remembered that conversation. However, unlike the other marginalized creatives, she openly stated she wanted to make a lot of money for herself and for the sake of her family.

In this regard, she was a bit of an outlier among the young creatives I knew, although certainly many young creatives in China share her desire for material wealth.

Yuki's and Wilson's quest for meaning, and Wilson's and Lily's concern for "vulnerable groups," were not unique among my participants. In the post-Tiananmen years, the party-state has engaged in a "compromise legitimacy" whereby rising living standards and material prosperity have been guaranteed in exchange for political rights (although China's economic problems prior to and especially since COVID-19 have troubled this "agreement"). However, despite (or perhaps because of) the abundance of creature comforts of every kind, many have noted a spiritual vacuum in China and a sense that, although the government seeks to cultivate a "socialist spiritual civilization" and the "Chinese Dream," there is a moral crisis afoot or a loss of purpose. These sentiments are manifest in the "psychologization of society" (Yang 2014b, 2015) and the popularization of positive psychology that blends Confucianism, Chinese Buddhism, and Daoism (Y. Zhang 2018; Zhang 2015, 2017, 2020), as discussed in the book's introduction. Millions have also sought meaning through Christianity in an officially atheist country (Yang 2011). Wilson actually did become a Christian several years after I met him. Instead of formal religion, however, most of the young creatives I knew sought meaning through tying their aesthetic endeavors to an ethical imperative.[81] Although most said they were not interested in politics, they were very concerned about social issues and how to make the world a better place, even if they as individuals actually had little power to do so given their position in Chinese society.

Conclusion

The Chinese state's embrace of innovation, and the top-down discourse that *anyone* can and should be creative, innovative, and entrepreneurial, in many ways can be understood as a more recent technology of governance, heightened in neo/non-liberal China, and made manifest in intensive efforts at cultivating creativity and promoting innovation as a form of value and human capital. However, to say that state discourses and policies are the main driver of these individual and societal transformations is to ignore people's own understanding of their embodied experiences as they are lived on the ground and give rise to desires and aspirations. In this chapter, I have shown how marginalized young creatives in Beijing exist within these discourses yet also try to stand outside of them through framing their creative pursuits in terms of something that moves them and as a search for meaning. They cultivated their sense of aesthetics as a way to gain the kind of education, knowledge, and skills that they found valuable as individuals. For them, aesthetic cultivation was a passion, an ethical practice, and a means of voice and self-expression. Their use of social media cannot be separated from such processes nor from the manner in which they cultivated and curated the self. Distancing themselves from dominant norms, such as eschewing rampant commercialism (except for Xiao Sun), was a way simultaneously to value themselves and map their own personal aesthetics into this terrain even if in some ways it was inarticulable—it was something they felt and to which they were drawn. Over time, it became evident that in pursuing knowledge and personal aesthetics, and in doing so cultivating the ethical self, the endpoint (Foucault's ethical *telos*) for these young creatives was arriving at a stage where a personal aesthetic became one's own. Although individualism in China is sometimes understood as selfishness, these young creatives embraced personal aesthetics as a path to create meaning and the betterment of themselves—intellectually, spiritually, morally—and by extension Chinese society through the ethical choices they made.

When I first met all of these young creatives, they accepted uncertainty and unstable financial prospects to pursue creativity, some working for jobs for a few months at a time to save enough money to pursue their dream, whether it be dancing, documentary filmmaking, or design. A reasonable question is, thus: how sustainable is this way of living in the long term? There is no simple answer to this question, because although the young creatives shared a similar class position, and they were all marginalized, they nonetheless had varying levels of economic, social, and cultural capital. For a while, Xiao Sun's job at the gallery enabled her to learn about art and the art trade. However, the last time I saw her, in 2019, the gallery where she had been employed had closed and her mentor had left Beijing. At a different gallery, Xiao Sun was often working ten hours a day at the beck and call of the gallery owner at all hours of the day and night via her mobile phone. As we kept in touch later via WeChat, she said she felt like a glorified event planner. She was unable to overcome her lack of formal education and cultural capital and was clearly situated in exploitative labor conditions.

Others left Beijing altogether. As already mentioned, Wilson had returned to Shandong province, yet he was able to continue his film-making because of his social networks. In 2017, Lily took me on a short walking tour around the *hutong* where her store was located to show me all the little shops that had been razed and buildings whose windows, formerly used by migrant vendors to sell everything from hot soy milk to toilet paper, had been cemented over. These changes had occurred as part of Xi Jinping's "beautification" campaign, which also resulted in the expelling of numerous migrant workers from the city, as I discussed in the book's introduction. Such repression and discarding those people and things that were no longer deemed useful, in this case on the pretext of creating beauty and order, was one more aspect of Xi's Chinese Dream. "Beijing is becoming less interesting," Lily said as we walked, adding that a lot of her creative friends were leaving. A year later she moved to a large city far from the nation's capital, stirred by hope of a less stifling environment in which to continue to pursue her own dream. In 2019 while we were both visiting Beijing, she seemed quite content. However, two years later she had moved back to Beijing, unable to resist the pull of finding something better.

2

Challenging Technosolutionism

Gender, Age, Micro-entrepreneurship, and
Relational Ties in the Countryside

It is a warm day in August 2019, and a Chinese colleague and I are in a taxi that has traveled about 85 kilometers from Jinan, the provincial capital of Shandong province. As we approach our destination, a city to the southeast, we see new high-rise apartment buildings towering on either side of the highway, a symbol of both the economic development that has been an ongoing phenomenon in the region and the local government's stepped up, and not completely welcome, urbanization plan.[1] We arrive at our hotel in the older part of the city (little changed since I last visited), check in, and then grab another taxi to Suan Village, where I had previously conducted research on the villagers' use of technology for micro-entrepreneurship.[2] As we pass through the township and arrive at the village, the visual changes are noticeable but less dramatic. The central government's increased rural revitalization efforts since 2017 have transformed what were previously dirt paths into paved walkways, graced on either side by flowers and vegetable patches. Small garbage dumpsters are also situated around the village, a welcome remedy to the intermittent trash piles that had formerly served that function.[3] My colleague and I exit the taxi, and as we head to our host's home, I see a familiar face. Approaching us on her three-wheeled motorized cart is Ms. Fang, whom I had met on my first trip to the village in 2011.[4] She is making her afternoon round of selling homemade tofu to the villagers. Her face lights up as she recognizes me, and we exchange a warm greeting. She offers to give us a ride to our host's house, and although the distance is quite short, we climb in.

In subsequent conversations with Ms. Fang and her husband, I learn that only a couple of months earlier they had begun renting a large

Figure 2.1. Ms. Fang making her rounds selling tofu and *mantou*. Photograph by the author.

compound on the main street. In addition to tofu, they had started making *mantou* (馒头), the wheat-based steamed buns that are so ubiquitous in northern China. The compound has eight rooms and is much larger than the village home they own, which sits empty. Ms. Fang also shows me the domestic-brand smartphone she purchased the previous year but says it hasn't changed her business much, nor have the printouts of Alipay and WeChat QR merchant codes she carries, because few people use them. Just as she has done for over a decade, Ms. Fang rises daily at 3:00am to prepare the tofu that she will sell later during the morning and afternoon as she rides around the village, a prerecorded male voice announcing her arrival through an electronic megaphone connected to the battery of her cart (see figure 2.1). As we talk, her husband receives frequent WeChat notifications. Some are from the village WeChat group, but Ms. Fang isn't a member. "I'm too busy. I have no time for that," she says, but she has joined the group set up by her young son's teachers, affirming her role as caregiver in the family.[5]

The scene described above is an edited excerpt from my fieldnotes after my first day back in Suan Village. The smartphone, QR merchant

codes, and WeChat groups mentioned by Ms. Fang are now ubiquitous in villages all over China. Like the paved walkways and vegetable
gardens, they denote gradual but important changes to village life, as
opposed to the dramatic economic and cultural transformations that
are supposed to have occurred as a result of a decades-long govern-
ment- and corporate-led rural informatization drive.[6] In the villages,
most "older" (i.e., over age 35) residents have incorporated smartphones
and limited social media use into previous patterns of work and family life, melding them with longstanding norms of trust, reciprocity,
and obligation in social relationships. Still, in neo/non-liberal China,
the government's call to be innovative and entrepreneurial, with digital
technology as a central component, is meant to summon all Chinese
citizens, including those in rural areas (A. Li 2017; B. Li 2017). Such
discourses often simultaneously acknowledge and elide regional economic, infrastructural, and technological asymmetries that have resulted from previous state development policies.

This chapter examines how small family businesses run by entrepreneurial couples engage with digital technology for micro-
entrepreneurship in the Shandong countryside. I situate the practices
of older micro-entrepreneurs like Ms. Fang within earlier efforts by
younger couples in order to show continuities and transformations.[7]
Shandong is one of China's wealthiest provinces, with the nation's third
largest provincial GDP per capita (National Bureau of Statistics 2021).
Even so, as one travels further away from the coast and from major cities, the uneven development that plagues much of China's countryside
is visible.[8] Based on fieldwork in villages during two separate periods
between 2011 and 2019 among two distinct types of entrepreneurial
couples, I ask, how have top-down discourses and policies that connect
technology to rural economic transformation manifested over time in
the villages and to what end? How is digital technology use for micro-
entrepreneurship articulated to the affective bonds and interpersonal
ethical norms that inhere in villagers' familial and social relationships?
How are gender, age, and migration experience connected to the use of
different platforms and devices, and what are the implications, especially
with regards to women's personal agency and economic empowerment?[9]
In answering these questions, I focus in particular on the consequences
for women because they were most often the ones running the daily

operations of the business (in some cases their husbands were not involved at all) even though patriarchal norms could obscure or devalue
their labor. Rural women, particularly older women, tend to be more
marginalized, but they still aspire to use technology in various ways to
contribute economically. I also pay attention to how the demonstration
of proper feeling and emotion (*ganqing* 感情) was constitutive of villagers' business interactions with fellow villagers. As well, I highlight how
at times the violation of the norms of *renqing* (人情), or "a system of
ethics that guides and regulates one's behavior" (Yan 1996, 145), emerged
in some of these interactions and in certain entrepreneurial practices,
which in turn shaped outcomes.[10]

I argue that regardless of background, villagers seek to use technology
in their businesses in ways they deem appropriate for their circumstances,
which, when only viewed through a narrow discourse of what "success"
or "talent" look like, invariably are overlooked or deemed to fall short.
Moreover, what is considered appropriate or suitable for one individual
might not be for another.[11] I thus show that women's degree of success
does not depend only on their own experiences and aptitudes, nor their
perceived role primarily as domestic caregivers. Honing in on how ethical norms are upheld or violated in the context of technology use, and
how these too are influenced by age and gender hierarchies, generates a
holistic understanding of the outcomes of villagers' business endeavors.
Ultimately, I argue that although women's economic contributions could
be masked or constrained, through engaging with technology and adopting various tactics, both younger and older women do carve out spaces for
agency and achieve some degree of empowerment in the family through
their entrepreneurial efforts. My findings are situated within and add nu-
ance to two dominant discourses about technology access and diffusion
in rural China: first, the insistence by the government, NGOs, and corporations that technology is the driving force of positive economic and
cultural change in the countryside, and second, a global discourse, which
is localized in China, that technology leads to empowerment, especially
for marginalized women (Aliresearch 2019, 2022; Hafkin and Huyer 2006;
Hossain and Beresford 2012; Mumporeze and Prieler 2017).

In what follows, after discussing the context for the research, I provide background on how state rural development efforts in the eighties

and nineties resulted in significant economic shifts that transformed gender relationships, family structures, and villagers' social ties, which led to what some called a "moral crisis." The remainder of the chapter is organized around two periods of fieldwork. The first took place towards the end of the informatization efforts linked to the "New Socialist Countryside" policies of the Hu Jintao-Wen Jiabao era and the beginning of Xi Jinping's rise to power and focuses on young, entrepreneurial couples who sought to take advantage of China's rural digital boom in relatively uncharted territory. This historical analysis provides context for comparison with the later fieldwork, carried out in the neo/non-liberal milieu characterized by state rural revitalization policies that intersected with corporate efforts to further e-commerce, the platform economy, and digital innovation in the countryside. These policies and plans were accompanied by discourses that emphasized the positive transformations brought by these processes. This section examines how middle-aged women, who were usually one half of an entrepreneurial couple, engaged with technology in their businesses. Together, these analyses add to understandings of the role of technology in personal and/or economic transformation in rural China through not only considering structural constraints, spatial dynamics, and age and gender relational patterns, but also ethical obligations, feelings, affective labor, and notions of intimacy and trust.

Finally, my definition of micro-entrepreneurship is necessarily broad, as I include what have been called "domestic sidelines" or the "courtyard economy," exemplified in Ms. Fang's production of tofu and *mantou*. As well, my account of the use of various technologies, such as the sewing machine, is expansive, in line with feminist understandings of labor and technology (Wajcman 2004, 2010). My analysis also considers factors that might seem tangential to entrepreneurship, such as the nuances of familial obligations, because in rural China, historically the bonds of family and kinship have run extremely deep and form the basis for one's notion of self and one's livelihood.[12] In the countryside, the blurring of work and family life, so lamented in contemporary research on digital media in western contexts (Gregg 2011; Wajcman, Bittman, and Brown 2008), has been a constant for centuries.[13]

Economic Reform, Gendered Transformations, and "Moral Disintegration"

China has undergone a dizzying array of economic, social, political, and cultural transformations over the last several decades, and perhaps these changes are no more dramatic than in the countryside, where they have affected labor dynamics, family structures, gender constructs, communal relationships, notions of ethics, and more (Jin 2011). Although the early years of China's reform and opening policy brought rural residents tangible material benefits after the agricultural communes were dissolved and the household responsibility system was implemented, this period was relatively short-lived.[14] By the early nineties, the government had shifted its development efforts to cities located primarily in the eastern coastal regions. In subsequent decades, peasant farmers struggled to make ends meet even as many tried to operate small businesses, do various side jobs, or find part- or full-time work in local factories that emerged as a result of rural industrialization (Bossen 2002; Croll 1987, 1995; Song 2015). Faced with economic hardship, numerous rural adults left agriculture altogether, becoming part of a growing number of rural-to-urban labor migrants.[15] Indeed, in many provinces, out migration and remittances sent back home became the primary means of supporting those "left behind": children, middle-aged women, and the elderly (Ye 2011).

The economic reforms have had particularly gendered consequences in rural areas. Although rigid gender binaries previously had been somewhat loosened, as gendered divisions of labor shifted, so too did the value of certain types of work. Such perceptions of value are based on centuries-old binaries whereby "masculine" and "feminine" are mapped onto notions of outer and inner, skilled and unskilled, and heavy and light work, respectively (Bray 1997; Jacka 1997). Jobs located in the household's "inner" realm, such as weaving, embroidery, or food preparation, have historically been seen as "women's work." In the 1980s, the All-China Women's Federation (ACWF) promoted the "courtyard economy," as it came to be called, as a way for women to contribute economically by doing what had been called "domestic sidelines." Because these blended in with women's household domestic tasks, they were not always viewed as labor.[16] Moreover, with most able-bodied male adults and young women having left villages as labor migrants, middle-aged women took on much of the responsibility for farming, resulting in the feminization of agriculture, now viewed as "inner" work (e.g., in the village) and thus less valued (Bossen 2002; Gao 1994; Jacka 1997).[17] These "left behind" women became discursively constructed in official and popular discourse as "vulnerable" and/or "victims," even as top-down development projects to "empower" them often stripped them of their agency (Jacka 2014; Jacka and Sargeson 2011).

Over time, economic reform and rural-to-urban migration not only shifted gender relations and notions of labor, but also resulted in the breakup of tight-knit communities formerly based on kinship, shared history, and mutual obligation. Such transformations, along with urbanization, consumerism, and of course, technology diffusion, are said to have completely unraveled the moral fabric of the countryside. Although alarms have been raised for years about the decline in ethical behavior and/or the loss of a moral compass among the general populace, such critique has been leveled in particular at rural residents. Some scholars have argued that traditional notions of filial piety (*xiao* 孝) and true human feeling, already damaged by the chaos of the Mao era, have been gutted over the last several decades.[18] In this view, the ensuing emphasis on wealth, consumption, and competition has led to villagers constructing arbitrary sets of practices, or an "immoral politics," in negotiating the challenges of everyday life (Liu 2000). Yan Yunxiang (2003, 2010) argues that while the embrace of individualization by younger rural generations has been liberating for them as they have gained more autonomy, especially in marital choice and private life, this shift has been accompanied by selfishness and lack of trust.

Less pessimistic observers acknowledge that contemporary notions of morality and ethics might be more contested than in the past, yet they note that ethical obligations still inhere in rural family bonds and kinship networks (Jacka 2014; McDonald 2016). For example, studying a village in southern China, Oxfeld (2010) uses the term *liangxin* (良心), or "conscience," as a means of unpacking how discourses about being virtuous and fulfilling one's moral obligations are embedded in memory and materialized in daily life among family members and in villagers' interactions based on reciprocity.[19] Steinmüller (2013), researching in a village in central China, focuses on *li* (礼), or ritual, to argue for the importance of what he calls "everyday ethics," or everyday activities that involve creative and reflective engagement with various moral frameworks.[20]

Oxfeld's (2010) and Steinmüller's (2013) careful ethnographic work recognizes both the durability and flexibility of Chinese relational ethics as they are embedded in shifting norms of familial and social relationships. Although both provide nuanced accounts of the interconnections between gender, economic livelihood, and "moral discourse" (Oxfeld) or "everyday ethics" (Steinmüller), the role of digital technology in such shifts is not addressed. Based on fieldwork in a rural town in Shandong province, McDonald (2016) shows how social media became the site for the expression of moral frameworks, for example through his informants' posts of images depicting idealized family life, their concerns around trust in online relationships with strangers, and their "moral accumulation" through collecting credits on various platforms.[21]

I build on this this body of scholarship through considering the entanglement of gender, age, migration experience, technology, emotion, and interpersonal ethics, as these manifest in villagers' social relationships and entrepreneurial endeavors. As Andrew Kipnis (1997, 10) has argued, *ganqing* constitutes "the individual and the social," and as such its expression is often performative in that displaying certain emotions, through actions as much as words, conveys one's care and concern for those in one's social networks, or *guanxiwang*.[22] Because *renqing* is composed of "rational calculation, moral obligation, and emotional attachment" (Yan 1996, 146), it is a useful analytic for understanding villagers' entwined business and social interactions, especially when ultimately personal interest was at stake.[23] Although there has been debate over the last several years regarding whether the importance of *guanxi* would fade with increasing marketization, much research has revealed the continuing importance of the intersection of *guanxi, renqing,* and *ganqing* in different contexts (Bian 2018; Yang 2002), including various forms of platform labor (Tang 2023; Peng and Wang 2021).[24]

The New Socialist Countryside and the Challenges of Connection

In 2011, during my first visit to rural Shandong, one of my hosts took me to visit his uncle, Mr. Liu, who lived in Zhu village, the most remote village of the six I visited. As a young man, Mr. Liu had done a variety of jobs as a migrant worker in different locales. When I met him, he was 38 and was married with a 16-year-old son. He had been a successful pig farmer for several years after struggling, and failing, to raise sheep. As we chatted, the conversation turned to a brand-new Lenovo computer sitting prominently on a desk in the living room. The computer was a gift to Mr. Liu's son, a reward for passing the high school entrance exam, yet because there were no computer shops in the village or nearby town, Mr. Liu had bought the computer with the help of a relative who had an internet café in the closest city. Although Mr. Liu's son would soon be going to boarding school, the computer would remain in the home. "He can use it when he's here on weekends," said Mr. Liu. "I want to use it to learn new techniques [related to raising pigs] and get daily prices and other things."[25] However, none of these activities was possible without an internet connection, and at the time of my visit, the family had already been waiting weeks for someone to hook up the broadband, and they were still waiting when I left.[26]

The fact that the Liu's remote village had broadband was a result of the Hu Jintao-Wen Jiabao leadership's (2002–2012) emphasis on "people-centered" development and "sustainable growth" (Qiang et al. 2009). By the beginning of the twenty-first century, the countryside had become an urgent policy issue because economic reform had not only upset traditional family and interpersonal relationships and brought the perceived moral decay discussed above, but also created alarming urban-rural and interprovincial inequality (Fan 2006).[27] The state's proposed remedy was to urbanize rural areas and launch a campaign to build a "New Socialist Countryside" through, among other things, greater investment in agriculture and rural transportation infrastructure (Fan 2006; Wen 2006).[28] Informatization was also seen as key, with attention placed on landline expansion followed by substantial diffusion of mobile phones and broadband internet to the countryside.[29]

At this time, telecommunication companies, both to comply with state policy and to tap into new markets, offered numerous promotional packages to encourage rural residents to purchase smartphones, computers, and internet connections, as the Liu family had done.[30] However, the Liu's long wait for broadband connectivity was not unique. Despite progress made in rural areas, many policies were critiqued for inefficiency, lack of accountability and sustainability, and problems with coordinating strategy and organization between different governmental levels, thus leading to the neglect of some rural areas and oversaturation in others (Liu 2012; Ting and Yi 2012; Xia 2010). These issues were compounded when there was little support from the government (Ma and Li 2011). By the early 2010s, in the wake of the 2008 global financial crisis, the state was promoting e-commerce, discussed in more detail later in the chapter, as a way for rural residents to better their lives. However, Mr. Liu's experience calls attention to the challenges faced by rural residents who were seeking to use technology to enhance their economic opportunities in the most basic way and how digital inequality in rural China was, and still is, linked to age, class, and differences in village type (Guo and Chen 2011; Shi and Ai 2013). It also reveals why, several years into the New Socialist Countryside policy, there were still many questions about how and if communication technology was connected to improving rural livelihoods.

The Early Promises and Perils of Technology and Micro-entrepreneurship

While I was in the villages in 2011 and 2013, it was common to find young married couples in which one or both members had returned after gaining new ideas, experiences, and financial resources through labor migration and then started small businesses, just as others had done before them (Mohapatra, Rozelle, and Goodhue 2007; Murphy 2002). These couples were unique, however, in that their aspiration was to provide communication technology devices and services where none or very little had existed before. In Yu village, for example, Wang Fan and her husband, who were both 29 when I met them in 2011, had opened an internet café two years earlier. It was not far from a large seafood processing plant, so they hoped to attract the workers after their shifts ended. Across the street was a small mobile phone shop that Li Mei (also 29) and her husband had opened for the same reason. Li Mei felt that while working outside the village she had learned skills, such as marketing, that she could use in her business, which was doing well. "There is a lot of competition," she said, but she maintained customers "by being honest."[31] To cater to the workers, Li Mei sold domestic brands as well as some knockoffs (of foreign brand phones). In Suan Village, Xiao Sui and her husband Liu Dong operated a store that sold low-end mobile phones, computers, satellite dishes, and other accessories and offered services such as repairs, swapping out batteries, and the like. Unpacking the experiences of these couples, especially those of the women, in these fledgling enterprises illuminates how well-established connections between gender, technology, and affective labor were simultaneously maintained and reconfigured as the couples and their fellow villagers tried to negotiate an emergent technological terrain. It also reveals the entanglement of feelings and ordinary ethics, or "the routine, everyday and highly personal aspects of ethics" (Stafford 2013, 6), in which at times the "gradual accumulation of personalized obligations," or *renqing* (Bian 2018, 604), enabled and constrained micro-entrepreneurship.

In rural China, these family business ventures would conventionally have been gendered masculine, due not only to gendered notions of technology but also, in the case of the internet café, to government discourse.[32] However, the women were the ones who actually "manned" the cafés and shops because their husbands were frequently occupied elsewhere. These "outer" spaces were then folded into the "inner" domestic sphere of the women and were used in ways traditionally regarded as feminine. For example, during the late morning on a summer day while I chatted with Wang Fan in the internet café, some of the computers were being used by middle school-aged youth, and a small group of young children was playing with Wang Fan's daughter in a corner. Across the street, Li Mei's mobile phone shop had toys and a cot in the middle of the floor (see figure 2.2). Similarly, in Suan Village, when there were no customers and the weather permitted, Xiao Sui spent most of her time on the dirt road outside her shop. She said she enjoyed chatting with other village women and watching the children play after school. Despite this merging of inner/outer, gendered norms remained relatively stable, as women could fulfill their expected duties as wives and mothers while contributing to the financial well-being of the household, and in the case of the internet café, providing learning opportunities for the next generation.

The women's interactions with customers, infused with *ganqing*, or good feeling, also reveal the ethical and affective dimensions of their entrepreneurial efforts. Although the shops and cafés functioned in some

Figure 2.2. Li Mei's mobile phone shop. Photograph by the author.

ways as informal daycare centers, Wang Fan would never have thought about charging the parents of the children for their use of the computers because they were all fellow villagers (*laoxiang* 老乡). Her affective labor was understood as a reciprocal obligation that villagers would return at some indefinite point.[33] Her relationship with factory workers, however, was completely different. Although Wang Fan knew some of them, others were from different villages, and thus they were further removed in her social network. Their time at the internet café was strictly a business transaction. Her only ethical obligation to them was to deliver the service they paid for. When other internet cafés opened, creating fierce competition, customers in turn did not necessarily have loyalty; rather, they went to whichever café had the cheapest price. In contrast, in Suan Village, which was smaller and more remote, Xiao Sui knew her customers much more intimately. For her, opening the doors at 5:00am so that villagers could charge their phones before going out to the fields was as common as watching their children play in the road.

Such ordinary ethical obligation did not necessarily always yield an equal return, especially with those further removed from one's closer ties. The experience of a young couple in Zhu Village exemplifies how

personal and business interests can clash in the context of digital inequality and illuminates the challenges villagers faced in their entrepreneurial efforts. Lin Ying and Hu Yang met in Shenzhen while they
were both migrant workers.[34] After they got married, they returned to
Hu Yang's village and opened a store that sold computers and computer
components. However, in their small village their business encountered
some unexpected problems. First, sales were slower than they had anticipated because the young adults who would have been their prime
customers continued leaving the village to labor elsewhere. When they
returned for Spring Festival, they often brought mobile phones, and in
some cases computers, as gifts for family members. Second, Hu Yang
and Lin Ying had to keep the price low for their customers, who had
limited financial resources. When they sold a computer, they made only
a small profit, which meant that Hu Yang also worked in the nearby rock
quarry. Third, their relational ties obligated them to provide services
they otherwise would not have. As Lin Ying related:

> When we sold a computer, we had to travel fairly far away to the cus-
> tomer's home. We would set everything up, show them how to use the
> computer, and leave. Then we would get a call the next day that some-
> thing was wrong. So, Hu Yang or sometimes both of us would go back
> to the home and answer the customer's questions or remind them how
> to do something, but nothing was actually wrong with the computer. We
> would leave, but then we would get called again. We'd return to the house
> and find there was no problem. . . . and on and on. We had to make so
> many useless service calls because rural customers really don't know how
> to use computers.[35]

Given the nature of village life, where reciprocal obligation, or *renqing*, is embedded into even such loose relational ties, Lin Ying and Hu
Yang felt they could not say no to these calls, which cost them time and
money, because that would go against social norms and might jeopar-
dize the reputation of their business. Making the calls could also help
them accumulate some sort of favor that they could call on in the future.
Although it was true that their customers had limited technical skills,
Lin Ying also implied that these numerous calls were a tactic (intentional or not) for the villagers to get some free training, but she couldn't

say anything that might insult them or make them lose face (*mianzi* 面子). Like all villagers, Lin Ying and Hu Yang had to keep good *renyuan* (人缘), which, as Hui Zhang (2013) has argued, is not just about mutual obligation, but also maintaining personal popularity so that others do not get envious of one's fortune or success. Yet, it was clear during our conversations that even though Lin Ying masked her irritation, she felt ethical norms had been breached, which is usually the time when ethics are thought about consciously (Lambek 2010). In the end, it was easier for them to close their business, using the excuse of Lin Ying's health. Instead, they planned to open a convenience store, which Lin Ying would run, a decision that reaffirmed conventional gendered divisions of "inner" and "outer," and "light" and "heavy," labor.

Eventually, all of these businesses closed for several reasons. In the case of their mobile phone shop, Liu Dong and Xiao Sui found that as the village population continued to decrease—a result of the government's urbanization plan and young people's continued out migration to towns and cities—they had fewer and fewer customers. Wang Fan's internet café closed not due to the competition from other internet cafés. Rather, villagers were able to go online on their own due to the subsequent diffusion of high-speed internet and low-cost smartphones.[36] In contrast, Lin Ying and Hu Yang were ahead of the computer wave, so to speak, and the lack of infrastructure at the time, combined with their rural customers' limited digital knowledge, led to the demise of their business only a year after they had opened it. Their situation also exemplifies how rational calculation and moral obligation were embedded in competing interests: Lin Ying and Hu Yang's hope to make money in the process of providing a service conflicted with their fellow villagers' desire to learn new skills.

Did these couples' engagement with technology as a means of entrepreneurship early on create a space for agency, especially for the women? On the one hand, the husbands took the lead in decisions regarding which merchandise was stocked. They also had the skills to repair phones and fix and assemble computers. Furthermore, because it was impossible for these couples to make a living solely through their businesses, the men had "outer" jobs considered skilled and heavy—Wang Fan's and Li Mei's husbands did construction, Liu Dong worked in a steel factory, and Hu Yang labored in a rock quarry. Still, as noted earlier, as these businesses were integrated into existing gender relations, they also opened up new possibilities for reconfiguring the relationship between gender and technology. This reconfiguration occurred through women's physical presence in places gendered masculine and through knowledge they gained through informal learning. Still, upsetting gendered relations of technology can simultaneously reproduce gender-typed work, in this case women performing unremunerated caregiving. It can also be temporary, as in the example of Lin Ying. Thus, as has been found with other women returnees, the women gained a limited degree of status and autonomy, which led to their exercising slightly more agency than non-migrant rural women (Fan 2004, 2008; Murphy 2002; Song 2015; N. Zhang 2013).[37]

Searching for Rural E-Commerce in Neo/non-liberal China

After the first period of fieldwork, I was aware that many changes had happened in the Chinese countryside. When I returned to the villages in the summer of 2019, I was eager to observe in person the effects of policies aimed at improving rural areas since I had last visited. In 2015, as part of the state's Internet Plus plan, poverty alleviation was more tightly linked to the use of digital technology and the development of e-commerce in rural areas.[38] The Chinese government had promoted rural e-commerce as a means of solving the entrenched economic inequality in the countryside since the mid-2000s. However, the connections between e-commerce, tech innovation, and poverty alleviation steadily increased under Xi Jinping.[39] In neo/non-liberal China, e-commerce fits seamlessly with calls starting in the mid-2010s for "mass entrepreneurship" and "mass innovation," which were intended to hail not only the urban, educated middle class but also rural residents discursively constructed as "lacking talent" and of low "quality" (*suzhi* 素质) into the subject position of innovator/entrepreneur. Most policymakers continue to hold an unwavering belief that rural e-commerce will enhance China's information economy in general, which has contributed significantly to China's GDP, and that it will develop rural areas specifically, through increasing incomes and spurring more consumption (A. Li 2017; You, Ren, and Zhang 2017). After Xi's 2017 call for greater rural revitalization efforts, the cultivation of younger, skilled "new farmers" (*xinnongmin* 新农民) was also emphasized (see B. Li 2017).[40]

Just as telecommunication companies such as China Mobile had expanded to the countryside after experiencing market saturation in urban areas in the early 2010s, later so too did social media and e-commerce firms. Some, such as Alibaba, China's online e-commerce giant, viewed this shift not only as a source of new customers but also as a way to alleviate rural poverty. Alibaba thus framed revitalizing the countryside through e-commerce as a "moral imperative" (L. Zhang 2023, 104).[41] For a number of years, Alibaba led the way in this joining together of corporate and government development efforts, but Tencent, JD.com, Pinduoduo, and others also jumped on the e-commerce bandwagon (Sy 2015; L. Zhang 2023).

One of Alibaba's major efforts has been the creation of "Taobao Villages," defined as a village that generates at least 10 million RMB (US $1.6 million) annually from e-commerce or where there are at least 100 online stores or 10 percent of households doing e-commerce (Aliresearch 2014). Taobao Villages produce everything from electronics to clothes to toys to furniture, which are then sold through thousands of online stores using the Taobao platform. A few such villages were first designated in 2009. By August 2020, there were over 5,425 Taobao Villages and 1,756 Taobao towns in 28 provinces and autonomous regions, with one trillion *yuan* in sales (Aliresearch 2020b), and livestreaming had also become a key means of selling goods, especially due to the pandemic.[42] Taobao Villages are said to narrow the urban-rural income gap and create new job opportunities (Qi, Zheng, and Guo 2019). Not surprisingly, areas with younger, more educated villagers (with at least a high school education—usually returned migrant workers, but also college graduates who have gone back) are also more likely to prosper (Qi, Zheng, and Guo 2019; L. Zhang 2023). Former migrant laborers who open their own Taobao stores report pride in being their own boss (Lin, Xie, and Lü 2016; X. Wang 2013). However, Taobao Villages have been blamed for creating imitation, competition, and rivalry between villagers, leading to falling profits, and upsetting traditional village bonds (Lin, Xie, and Lü 2016; Lin 2023). Taobao Villages ushered in what Lin, Xie, and Lü (2016) call "an emerging hybrid rurality," bringing economic benefits but also changing the moral fiber of close-knit communities.

The positive reports from the government on communication technology diffusion to the countryside, and in particular the Taobao success stories touted by Aliresearch and state media (which often do not address the negative aspects mentioned above), are one part of the picture of how platforms have been integrated into and transformed rural life. The regions where e-commerce has thrived were already more prosperous due to earlier economic policies, so they could take advantage of established infrastructure, including roads, transport systems, and factories, as well as human capital (A. Li 2017; Zeng and Guo 2016).[43] However, infrastructure (or lack of) is only one aspect of the unequal starting points for e-commerce in the vast Chinese countryside. The large number of older people who reside there also means that not everyone easily participates in these assumed avenues for economic success.

During the second period of fieldwork, aside from the outward changes to the villages mentioned earlier, there were other noticeable differences. Inside people's homes, computers with internet access were now commonplace (even my host's elderly parents had Wi-Fi), and some residents had bought cars. Nonetheless, many of the other young entrepreneurial couples I had met during prior visits (in addition to those discussed earlier) had left and moved to the nearest town or city to make a better living. However, there were still a number of older entrepreneurial couples, where just as before, the women actually ran the business (despite calling it "ours"), and these women became my main informants. Most of the women were in their forties (a few were older) and had children, but very few had migration experience. They used smartphones and WeChat in some way for their business, in contrast to what I found in earlier fieldwork. Their homes also had computers with internet connections, which they used primarily for entertainment, in contrast to their husbands who read news and networked online. Below, I discuss ways that these women deployed digital platforms in their businesses, from minimally incorporating social media to divergent attempts at e-commerce. Like the young women documented above, as they aspired to participate in various modes of digital commerce, they faced numerous impediments—infrastructural, relational, financial, technological, etc. To deal with these challenges, the women used various strategies, some that existed in a gray area.

"Villagers Prefer Cash"

At first glance, it seemed that little had changed in villagers' technology use for micro-entrepreneurship compared to earlier visits, as many informants continued to run their businesses in fairly conventional ways even as rural e-commerce had taken off and high-tech experiments by younger, skilled "new farmers" were being carried out elsewhere (B. Li 2017; Wang 2020). As mentioned earlier, Ms. Fang was still driving around the village in her three-wheeled cart selling tofu and *mantou*, and her business remained steady.[44] She did not use WeChat to market her goods because her customers were fellow villagers. Regarding the printouts of WeChat and Alipay QR codes she carried around, Ms. Fang said no one ever paid that way except for a few younger residents.[45] Other women proudly displayed an array of WeChat and Alipay QR codes in their businesses but also told me they were rarely used.

In numerous government documents over the years, the rural populace, especially middle-aged and elderly women, have been seen as "lacking": culturally, educationally, technologically, and morally. Taking such discourse as truth, it would be logical to assume the villagers did not use the QR codes because they did not know how. In fact, one reason they eschewed usage had more to do with what they perceived as being appropriate. At a fundamental level, most older residents just did not see the need for digital payments. When asked about the QR codes in her small convenience store, Ms. Xu, who was 60 years old, said, "Just as face to face is better [compared to WeChat messaging and groups], villagers prefer cash. It's what we are used to."[46] An additional reason was that most daily purchases seemed too trivial to use a mobile phone for payment. Mr. Shen, who owned a television and satellite dish store, said, "Most people use cash, but if it's a big purchase like a television they'll use WeChat. But actually, more and more people are buying those kinds of goods through Taobao because it's cheaper, but the quality isn't as good."[47] Rather than a manifestation of "lack," the absence of digital payments revealed how relational networks and interpersonal interactions built on trust undergirded the flow of buying and selling goods in the village.

In further conversations, issues of trust became more prominent. Even though WeChat and Alipay send the payee a confirmation, villagers like to see their cash exchanges in their own hands. Fears about being cheated are not unfounded. Over the years, there have been a series of events that have broken down trust—crooked local officials involved in land grabs, factory owners knowingly releasing toxins into the environment, and individuals, including those involved in e-commerce, scamming people online, just to name a few. Indeed, the very rural revitalization efforts that brought digital media and e-commerce to the countryside have been rife with corruption as unscrupulous government officials and local businesspeople have used these as a means for personal profit rather than poverty alleviation (Boullenois 2020).[48]

Interestingly, the use of cash rather than digital payments—a result of the so-called backwardness of the rural populace—resulted in a possible mode of empowerment for the entrepreneurial women I met. Every one of them who had Alipay and WeChat QR codes said these were linked to their husband's mobile phone. The reason was that family bank accounts are almost always in the husband's name, regardless of who actually runs the day-to-day operations of the business. The customers who eschew electronic payments were thus unintentionally giving the women more control of the daily finances. That the use of something as seemingly trivial as a QR code also potentially reinstantiates gendered hierarchies is in many ways not surprising, given the history of technologies and gendered relations of power.[49] In this regard, the lack of use, which arose from relational norms and values of trust, benefitted women, at least in a small way. However, these same relational norms could disempower women, as I discuss later.

The Courtyard Workshop as Antidote to the World's Factory

Family workshops specializing in a particular item—textiles, shoes, leather goods—have been a mainstay of China's socialist market economy since the eighties (Zhang 2001; L. Zhang 2023). Indeed, women's involvement in weaving and textile work goes back centuries (Bray 1997). In rural Shandong, although I encountered small family workshops during earlier fieldwork, it was only in 2019 that I met women who were more actively involved in starting such businesses and using technology to facilitate them, as I illustrate with Ms. Gu's story.[50] At first glance, Ms. Gu's workshop appeared "traditional"; she did not have a website, there was no online selling, and constant expansion of market share was not the goal. However, understanding her and other women's motivations for participating in her workshop—as employer or employee—illuminates the shaping of the technological by relational norms. In other words, the women created an environment in which they could engage in productive and reproductive labor at least somewhat on their own terms.

Ms. Gu was in her early forties, married, and had two young adult children. In the spring of 2018, she and her husband converted their front courtyard, which sat off the main street of Suan Village, into a sewing workshop. Ms. Gu and her ten employees, women who all lived in Suan Village and surrounding villages, sewed lightweight white cotton gloves for workers at an electronics factory (see figure 2.3). The gloves are worn by workers to avoid leaving dust or fingerprints while handling the electronics. In an interview, Ms. Gu explained how she started this enterprise:

> In the past, I had worked at a large factory in Tai'an [a prefecture-level city about 80 kilometers from the village]. I knew there were factories there that did contract work with small workshops. [When I decided to start a business] I went there and approached one of the factory bosses. . . . He supplied me with the machines, fabric, and thread.[51]

Before deciding to work with this factory owner, Ms. Gu had considered his piecemeal rate compared to others and the volume he expected her to produce. Under their subsequent arrangement, he would call or send Ms. Gu a WeChat message when he had an order, and then schedule a delivery of material and thread to her. She messaged him when the order was almost finished, and then someone from the factory drove to the village to pick it up. Ms. Gu had built a relationship (*guanxi*) with this factory manager, but she was not interested in expanding her business with others. One reason was that building such trust took time, which is difficult when the other person is an "outsider" (not in the village).

The arrangement between Ms. Gu and the factory manager, based on economic obligation, differed from the relational obligations and affective ties among Ms. Gu and her employees. Like her, they all were

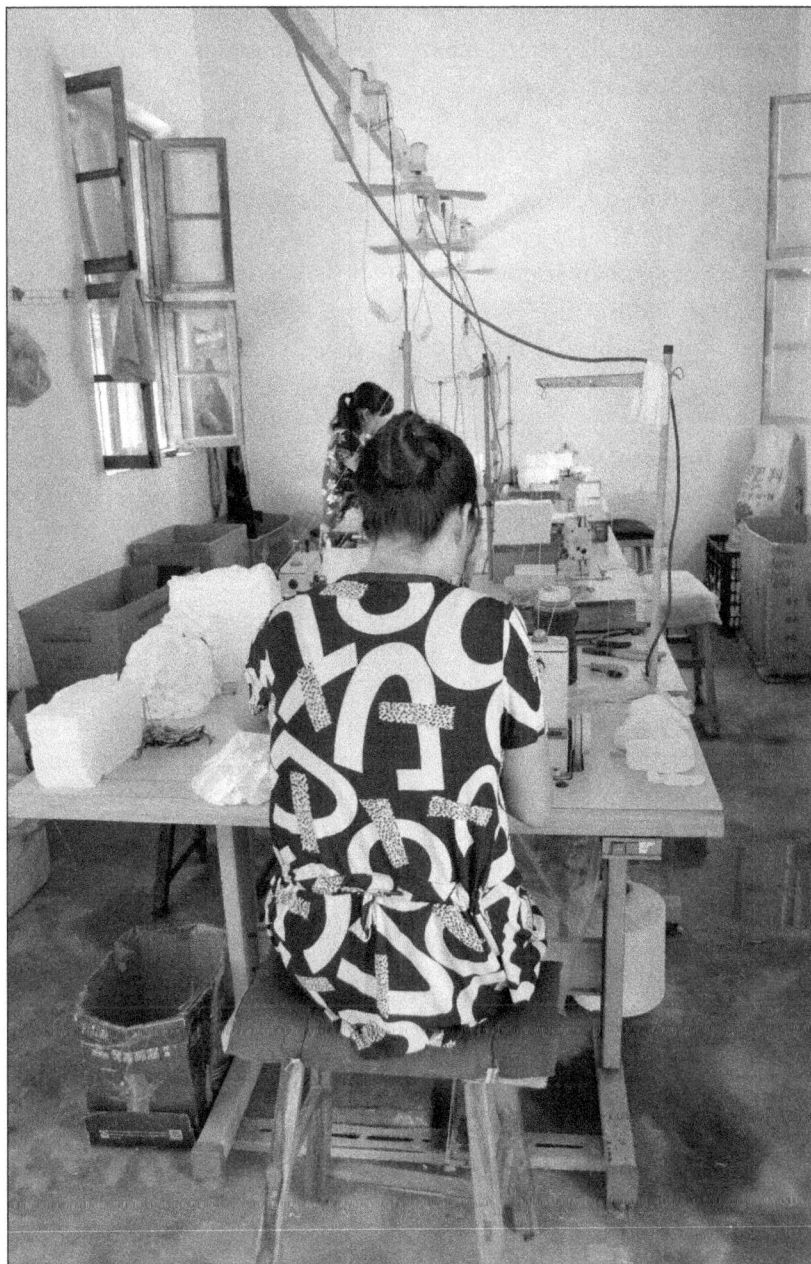

Figure 2.3. Ms. Gu's sewing workshop. Photograph by the author.

married with children, and the job allowed them the flexibility they needed to still be able to handle their domestic duties. Of her ten employees, six sewed alongside Ms. Gu in her workshop; the other three worked at home using sewing machines she had loaned them.[52] Some women only worked a half a day each day so that they could meet their elementary-school age children when they came home for lunch. With her employees, Ms. Gu had formed a WeChat group so that she could communicate with the women about their schedules, when she received new orders, and the like.[53] In many ways, the sewing workshop was an intimate space, where maintaining good feelings was important, and it allowed the women to share news, gossip, and knowledge, confirming Bray's (2008, 152) assertion that "the lens of intimacy reveals how technological choices and practices mediate between the emotional and material micro-dynamics of family life and the broader imperatives of livelihood, political economy and citizenship."

During previous fieldwork, I had found that men were often the ones who set up businesses. In this regard, Ms. Gu stood out in going to Tai'an and seeking out the factory on her own. Although her husband was not involved in this process, he did help initially by posting an announcement seeking workers in a WeChat group he had organized. I had also observed during earlier visits that women were excluded from the men's informal learning networks facilitated by technology (early on via QQ) (see Wallis 2015), yet Ms. Gu was a member of her husband's WeChat group. It was helpful to her because members shared about textiles, garment making, and other information. However, it was also somewhat intimidating. "There are many bosses in this group," Ms. Gu said, implying that although women were part of networks of which they were formerly kept out, her subordinate status left her lacking confidence to participate.

The work done by Ms. Gu did not fit the state's model of "modern" entrepreneurship, which in rural areas in neo/non-liberal China is often constitutive of more advanced uses of social media and/or an e-commerce platform. That she was doing contract work for a factory manager also does not fit narrow definitions of entrepreneurship. Nonetheless, the sewing workshop revealed Ms. Gu's acumen and skill in several realms, including sewing, negotiating business contracts, managing employees, accounting, working efficiently, and using technology—both the sewing machine and social media—not to mention doing all these tasks while tending to her own household. Within this assemblage, all of the women's skills and schedules were accommodated. As well, the types of exploitation, spatial control, and alienation faced by workers, mostly from rural areas, who labor in southern China's "world's factory," as detailed in the book's introduction (see also Pun 2005; Qiu 2016), and in textile production in urban migrant enclaves (Zhang 2001), appeared to be minimized.

Still, the women's sewing labor did not entail a reduction in their care work. Moreover, gendered power differentials attached to labor were not erased. Ms. Gu's husband had a job in a nearby textile factory doing what was discursively constructed as "skilled" work. However, Ms. Gu's presence in the WeChat group, which would not have happened earlier, and her relationship with the factory manager, where she had set at least some of the terms, represented a small shift in these power relations. Admittedly, her contract work with the factory could be viewed as yet one more example of the unequal relationship between labor and capital, especially since informal employment can disempower women (Boeri 2018). However, if we take Ms. Gu on her terms, we see that she was balancing her material and affective labor in a way she deemed suitable and in which she had some control.

Rural Taobao, Rural "Talent," and Reciprocity

Ms. Gu's experience represents the incremental steps that older village women have taken to use social media in small ways to accommodate the needs of their conventional businesses and fulfill their obligations of care work. Other women, especially if their children were grown, sought out opportunities to participate in what were at the time newer modes of technological entrepreneurship. I next discuss Ms. Jing and her engagement with Alibaba's "Rural Taobao" initiative. Like that of the young couples featured earlier, Ms. Jing's narrative once again draws needed attention to the challenges posed by scattered informatization, structural impediments, and the relational nature of village life, which encompasses gendered and aged constructions of power and obligation. These forces can disempower even the most industrious villagers.

In part as a recognition that relatively few villages have the infrastructure and human capital to actually become Taobao Villages, in

2014 Alibaba created Rural Taobao. Through its "A Thousand Counties and Ten Thousand Villages" project, Alibaba invested heavily to build an extensive network of service stations in villages and counties. The goal was to help farmers, stimulate economic development, and alleviate poverty in the countryside (A. Li 2017; Yu 2017). The service stations featured sample goods and "station chiefs" who were supposed to help rural residents browse products on flat-screen monitors and make purchases. Alibaba assisted with the transport of goods to the village stations, and the station chiefs were then expected to deliver goods to customers (Khanna et al. 2019). Alibaba hoped to expand Rural Taobao to more counties and villages so that rural residents could conveniently shop online (A. Li 2017). Rural Taobao was also supposed to enable farmers to sell their products online. In April 2016, Rural Taobao was in 16,500 villages in most regions of China, and Alibaba had a goal of reaching 100,000 villages between 2019 and 2021 (Aliresearch 2016). By 2017, there were more than 30,000 service centers in 700 counties (He 2017). Rural Taobao had significant government buy-in and support at the local and provincial levels (Khanna et al. 2019; A. Li 2017).

Just like Taobao Villages, Rural Taobao was initially heralded for bringing economic development (Aliresearch 2020a; He 2017; Zi 2019). However, there were early indications that it suffered from some of the same issues as Taobao Villages. By choosing station chiefs who owned businesses such as convenience stores, and paying them a commission on Rural Taobao purchases, Alibaba created a system in which the station chiefs were not motivated to help villagers buy online because it cut into their own businesses.[54] Even after Alibaba hired full-time station chiefs, only some areas prospered through involvement with Rural Taobao. Although buyers of goods seemed to benefit in some ways (Couture, Faber, Gu, and Liu 2021), local merchants complained that they were not able to sell their agricultural products effectively and that Rural Taobao created competition between local family businesses (Hap 2018). Yu (2017) argued that Rural Taobao was just a vehicle for Alibaba to sell things without really considering poverty alleviation and economic and societal transformation. Not surprisingly, in state rhetoric, a lack of "talent" was also blamed for disappointing results (Liao 2018). Eventually, Alibaba completely "upgraded" Rural Taobao and folded it into Tmall (its business-to-consumer platform), which resulted in further logistical problems and discontent among users and sellers (Li and Hao 2018).

Missing from these analyses of Rural Taobao's disappointing results are the gendered implications of a system that by default is linked to the relational ties that undergird e-commerce in rural areas, where family businesses are the norm. Ms. Jing's experience brings these issues to light. In her late forties when I met her, Ms. Jing had lived in Suan Village since her marriage to a local man in 1997. She was unique among the village women because she had grown up in a city and had an urban *hukou* (household registration), which she gave up upon marriage. Ms. Jing said that when she first moved to the village, the locals were skeptical about her ability to do physical labor. However, she was determined to show them that she could, and eventually they acknowledged her abilities. She had run a small convenience store since 2000, following in the footsteps of her mother and sister, who also had stores in a city.

A few years before I met her, Ms. Jing had become an agent for a Rural Taobao station chief, Lao Luo, a male relative living in the nearby township. She had helped him sell goods offered by Taobao to her fellow villagers. As she related:

> The way it worked was I would acquire some physical samples of goods, such as toilet paper. Customers could see the items, and I would tell them about the quality and the price. I also showed them other goods through the app on my phone. When a customer bought something, they paid me in cash and then I used Alipay to pay Lao Luo. He then ordered the products, and after they were delivered to his store, I had to go pick them up and bring them here. Then the customers would come to my store to get their purchases.[55]

Ms. Jing did really well at selling all kinds of goods—clothing, electronics, cooking supplies. However, for all of her skill and effort, she only made a small sum for each item she sold, no matter its price, and her annual income from Rural Taobao was only about 2,000 *yuan*.[56] She eventually stopped her involvement because it was too much work for too little of a return.[57] Also, more and more villagers started to download the Taobao app and use it on their own. They no longer needed her as a middleperson, just as Wang Fan's internet café was no longer needed when villagers were able to go online using their smartphones.

Ms. Jing's experience contrasted greatly with the enthusiastic discourse initially surrounding Rural Taobao and the subsequent reasons given in media reports and official documents for its disappointing results, which acknowledged structural issues but also blamed rural residents for their lack of knowledge. At the macro level, limited infrastructure was certainly a problem. Ms. Jing could not be a Rural Taobao station chief on her own because at the time there were no delivery companies that would service Suan Village. She had approached Alibaba about this issue, but Alibaba did not express interest, apparently because it would not have been profitable enough. Even in 2019, although villagers could receive packages from delivery companies, if they needed to physically mail a package, they still had to travel to the town ten kilometers away.

Probing deeper, however, reveals how gendered hierarchies and a violation of norms of *renqing*, or the demonstration of "the natural affect and feelings of attachments and obligations to other people" (Yang 1994, 68), also contributed to Ms. Jing's supposed failure. As relatives, Ms. Jing and Lao Luo were supposed to treat each other with trust and respect. On the one hand, Lao Luo was helping Ms. Jing because she did not have the ability to work for Rural Taobao on her own, but ideally they would both have benefitted from the arrangement. Technically, their collaboration defied Rural Taobao's rules because Ms. Jing was not certified as a station chief, yet as is often the case when the rules of the powerful do not accommodate local circumstances, which they have helped create, people find their own way of circumventing them without feeling that they have violated their own moral values. On the other hand, as a slightly older male relative, Lao Luo's status also meant that he could set the terms of their business relationship. Although Ms. Jing did not state this to me directly, as that would have communicated that she was ungrateful for his help, she implied that she had been taken advantage of by him. In other words, by helping Lao Luo, she was "*mai renqing*" (literally "selling *renqing*") in the hope that he would treat her equitably and return the favor in the future (if *renqing* in this instance is viewed like a raincheck). However, he disregarded this unspoken agreement (e.g., paying her a pittance instead of a percentage of the cost of the item sold), apparently without any shame, which should not occur when a relationship is built on trust and empathy (Sundararajan 2014). We could also say he lacked conscience (*liangxin*) because he didn't fulfill his ethical obligations.

Like Lin Ying, to get out of this situation, it was easier for Ms. Jing to create an excuse, in this case that villagers were buying things on their own and no longer needed her, than to state directly that she felt taken advantage of, which could make Lao Luo lose face. This way, amicable relations were maintained and could be called upon in the future if needed. Also, as with Lin Ying, Ms. Jing's free labor did provide her customers with informal training and knowledge, which Ms. Jing did not regret because she had demonstrated *renqing*, which makes one morally worthy or virtuous. This help also created an unspoken expectation that these customers would continue to shop at her store.

As Rural Taobao declined, Alibaba pushed forward with its online Taobao platform in rural areas, even enthusiastically declaring that women, including rural women and a surge in "retired moms" over 55, made up over half of its online entrepreneurs (Aliresearch 2019). However, scholarly studies on Taobao that focus on gender have found that women's participation in Taobao enterprises may empower some women economically while disempowering others, especially older village women (Liu 2020; L. Zhang 2023). Women's participation in Taobao has also not necessarily shifted traditional kinship patriarchal relations. Instead, some research has shown that in some ways it has reinforced them, especially for return migrant women who resume a subordinate role in the family (Lin, Xie, and Lü 2016; Yu and Cai 2019), while other research argues that women gain more bargaining power (Liu 2020); in both cases women still bear the burden for domestic responsibilities.[58] Ms. Jing's example similarly shows how someone who was well-educated and capable could not escape the patriarchal norms that subjugated her talent and skills to men's. Of course, structural constraints also stymied her entrepreneurial aspirations. Her only "e-commerce" when I left the village was to occasionally advertise one-off products on her WeChat. If she came across something she thought would be a good bargain for those in her local network, such as watches, she posted it on her WeChat Moments, where her close ties would be likely to trust her judgment.[59]

WeChat and "Gray" Commerce

During this period of fieldwork, eventually my host became concerned that I hadn't met anyone who engaged in what would conventionally be called e-commerce. She continued to put out queries around Suan Village and villages nearby, and after one particular WeChat message exchange, she declared, "I found one!" A fellow villager told her about a woman, Ms. Chen, who was doing quite well selling medicinal creams on WeChat. As I discuss below, Ms. Chen's story brings to light the way in which in newer modes of rural e-commerce, ethical obligations and reciprocity shift in accordance with greater relational distance, and it reveals even more so how certain practices exist in a moral gray area.

Like other businesswomen I met in Suan Village, Ms. Chen was in her mid-forties, married, and had two teenaged children. She had done various types of labor in the past, including working in a nearby agricultural processing factory. In 2017, she was contacted via WeChat by a woman she didn't know. At the time, according to Ms. Chen, "I was a housewife and wasn't doing anything and felt bored."[60] Although Ms. Chen tended to all the domestic chores and did the farming in fall and spring, her comment is indicative of the longstanding invisibility of rural women's labor and how it is undervalued even by the women themselves (Judd 1994). The woman tried adding Ms. Chen as a WeChat contact several times.[61] At first Ms. Chen refused, yet the woman persisted, and the third time, for reasons Ms. Chen did not explain, she accepted the request. This woman was an agent of Hibiscus Skincare Company, which manufactured herbal remedies for various skin conditions and bodily ailments. She pitched some of her products to Ms. Chen, who took the bait. "I had athlete's foot and found a cream to cure it. I used it, and it really worked," she said.[62] The woman then recruited Ms. Chen to sell the cream and other items.

When Ms. Chen was just starting out, she had to negotiate a bit of a relational quandary. She only had about 25 WeChat contacts (all friends and family), and she tried selling the creams to them, including via We-Chat. However, this mode of doing business was unsuccessful because her contacts quickly grew tired of her pitch. Also, Ms. Chen had violated relational norms because they had not indicated that they were interested in her products. As such, her actions could be seen as lying outside the typical rules of reciprocity in which one's voluntary action, such as a favor, could be reciprocated in the future.[63] Ms. Chen was like many other rural women at the time who were trying to engage in "micro-business" (*weishang* 微商, this term is also used for the seller), meaning they used a social media platform (often WeChat) to try to sell various items (Wang and Sandner 2019).[64] Ms. Chen eventually settled on a more conventional means of marketing to people in her village and surrounding villages. As we talked, she handed me a two-sided color brochure with pictures and descriptions of various products. She said she had recently printed 2,000 copies of the brochure, which she intended to distribute to friends and relatives as well as at the village market once a week, where such advertisements were more common.

Once Ms. Chen became an agent, there was an expectation that she would not only sell products but also recruit others into the business (the brochure had a QR code one could scan to receive information about becoming a national agent). The woman who enlisted Ms. Chen taught her how to add WeChat friends and strangers, such as by using the "shake" function.[65] During our interview, I asked Ms. Chen how many contacts she currently had, and she showed me her phone. There were over 4,000 people divided into around 30 groups (she said at one time she had had 80 groups but purged some of them). She also had over 10 sales agents beneath her, and together with her they sold 200–300 tubes of cream per month and made 2,000–3,000 *yuan*, with her taking a larger percentage of the profit and the downline agents splitting the rest. The more agents Ms. Chen recruited, the more money she could make. Later in our conversation, she showed me a 30-second video of her boss on Douyin. In the video, a middle-aged woman was quickly wrapping up some tubes of cream while touting the cream's healing properties. The price and further details are not mentioned. If a viewer wanted to find out more, they had to add this woman on WeChat. In other words, it was a typical clickbait advertisement.

The business venture in which Ms. Chen was involved in many ways blurred the line between multi-level marketing and a full-blown pyramid scheme, both of which are illegal in China and have drawn increasing government scrutiny. This method, with short videos and livestreams, is commonly used by companies that try to recruit rural women as *weishang*.[66] At the same time that Ms. Chen joined Hibiscus

Skincare as a salesperson, the founder and several members of one of China's biggest pyramid schemes, *Shanxinhui*, was arrested, and a year later he and nine of his staff were fined and given jail sentences.[67] What was striking about *Shanxinhui* was how it epitomized several facets of the neo/non-liberal milieu—tapping into state rhetoric on "poverty alleviation" and doing good while melding the promise of prospering financially with the broader cultural discourses around happiness that draw from Buddhist and other indigenous philosophies, as discussed in the book's introduction.[68] Other e-commerce businesses use such techniques as well, yet from Ms. Chen's description of her company and her boss's original recruitment of her, explicitly spiritual language was not used. However, there were some similar aspects: as evident from the brochure Ms. Chen gave me, Hibiscus Skincare used testimonials as a large part of their pitch, promising physical healing for everything from dermatitis to hormone imbalances to cataracts to spinal stenosis. The brochure used language and images meant to stir the potential customer to purchase the cream in order to heal. Of course, there was the added promise of profit if they became an agent.

It is worth noting that Ms. Chen's use of WeChat rather than Taobao was not necessarily due to the lack of village infrastructure or her own limited technical skills. It is likely she and the other agents were not allowed to use Taobao to sell the Hibiscus Skincare products because the company was skirting regulation and compliance with the National Medical Products Administration, which oversees food, drug, and medicine (including traditional Chinese medicine) production and safety. Taobao tries to ensure quality control through its suppliers and vendors. If standards are not met, however, and a factory still wants to sell goods, there are multiple workarounds that people can use. Selling on WeChat was one such way because it was less regulated as a selling platform.

In a sense, Ms. Chen presents an opposite case from that of the other women I met in the villages across all my trips. After temporarily disregarding the everyday ethical norms of interpersonal exchange in the village, she managed to find a means of economic empowerment, the details of which potentially violated laws, but (eventually) not necessarily ethics. Although operating in a legal gray area, Ms. Chen could justify her work because she claimed to truly believe in and have benefited from the products (and unlike a true Ponzi scheme, there was an actual product). Like the other women, her work schedule was flexible, perhaps the most flexible of all, and she could do the job while taking care of her family. Indeed, in her extended family she felt she had gained more respect because of the money she was earning even if some of her actions were most likely the topic of village gossip. This situation thus exemplifies Steinmüller's (2013) "communities of complicity" because those around her who were benefiting turned a blind eye to the potential illegality of the work (not to mention the unequal payment structure for those working down the chain).

Conclusion

For decades, the Chinese government has put faith in rural informatization as a means of spurring economic development, and by extension, the improvement of livelihoods in the Chinese countryside. Earlier, the emphasis was on expanding telecommunication infrastructure and providing access to computers, mobile phones, and the internet to rural residents. More recently, in concert with corporations like Alibaba, Tencent, and others, e-commerce in particular has been seen as a key driver of rural wealth creation and the elimination of rural poverty. Throughout both stages, there has been a persistent, often unquestioned assumption that technology will bring unilaterally positive outcomes to the regions of China that are often perceived as "backward." This technological solutionism, with innovation a constant buzzword in popular and official discourse in neo/non-liberal China, often glosses over social, cultural, and structural realities.

In this chapter, I examined two different periods in which married couples in rural Shandong integrated technology in small but important ways into their micro-entrepreneurship, with particular focus on how gender, age, everyday ethical practices, and affective relational ties are inseparable from these endeavors. The younger couples in the earlier fieldwork sought to stake out a claim in the New Socialist Countryside, in which the promises of mobile and internet diffusion created opportunities for technological entrepreneurship and new aspirations for how one might prosper. The trials the couples faced, which eventually became insurmountable, resulted from uneven technology diffusion (and then eventual saturation), underdeveloped infrastructure, and a customer base whose degree of technological skill and relational closeness or remoteness greatly influenced the success or failure of the couples' businesses. The middle-aged women in the second period of fieldwork encountered some of these same challenges. However, their experiences add further insight into how marginalized older individuals adopt technology and economic strategies in diverse and sometimes unexpected ways in accordance with their own material circumstances, skillsets, and values. In both periods, women did achieve some empowerment through gaining technological skills, reconfiguring spaces, and contributing economically to their households.

This chapter has generated two interconnected findings regarding technology, platforms, gender, age, and micro-entrepreneurship in the countryside. First, all of the research participants, men and women, sought to incorporate technology into their lives and businesses in ways they deemed appropriate. For women, this often meant that they seamlessly folded their "productive" work into their domestic responsibilities, which also meant that some of the women's business or technological skills could be overlooked. In the case of the young women, gendered hierarchies were simultaneously challenged and reinforced as the women took control of spaces deemed masculine. Most of the women made their workplaces (mobile phone shops, internet cafés) equal parts places for technology use/diffusion and care work, in the process gaining technology skills and, in some cases, providing opportunities for children to learn skills as well. Fellow villagers also benefited, even if their learning was not mutually beneficial for their informal instructors.

During the second period, I sought to highlight how older rural women, who are often maligned in popular discourse in China and left out of scholarly discussions, had their own way of integrating technology into their micro-entrepreneurship, even if some of these strategies might have seemed "traditional" or naïve to an outsider, especially when "innovation" becomes the marker of all "progress." Like the young women earlier, these women arranged space and time in accordance with both their business and familial obligations. Thus, like the work of other feminist scholars who have conducted research in rural China (Jacka 2014; Oreglia 2014; Wang and Sandner 2019), my analysis reveals that preconceived notions of older rural women's "lack" and "vulnerability" can hinder policymakers from understanding how rural entrepreneurial women negotiate various structures, discourses, platforms, power relations, and constraints that are constitutive of their lives.

Second, and closely related to the point above, gesturing to abstract notions of village women's lack of talent or their responsibility for domestic duties as factors constraining their economic success risks obscuring the nuances of how their engagement in micro-entrepreneurship is influenced by relational ties. *Ganqing* and *renqing*, at different times and to varying degrees, were constitutive elements of women's productive labor as it was articulated to technology use. Of course, all the villagers were embedded in social networks in which maintaining good feelings and upholding norms of reciprocity were important. However, when these were violated, even in ways that did not seem overly egregious, one strategy for gracefully exiting, thereby saving face for all involved and maintaining harmony, was to fall back on more conventional gendered divisions of labor, as in the case of Lin Ying and Hu Yang. Ms. Jing used this approach as well when her counterpart held more power due to age and gender and as such felt justified in putting his self-interest above hers. In contrast, some women were themselves the ones involved in questionable business practices, but because these practices existed in a gray area, they were considered acceptable by the woman and her close ties.

This chapter reveals that Yang's (1994, 110) assertion that "ethics and tactics coexist in tension and harmony" is still true in rural China. Regardless of their business endeavor, women were embedded in particular arrangements of ethical obligation and affective labor. Unpacking how they navigated these while using technology in ways they deemed appropriate, and the implications and outcomes of such, enables insight into their aspirations, agency, and incremental empowerment.

3

Domestic Workers, Virtual Voice, Care, and Community

In October 2017, while attending a Saturday activity sponsored by an NGO in Beijing that serves the needs of domestic workers, I sat with a group of women during the lunch break as they relaxed and chatted about their everyday lives. Some of the conversation was lighthearted, such as talking about the details of a new dish they had learned to cook or grabbing one's phone to play a funny video. Other topics were painful, however, as when Ms. Li shared how her previous employer had refused to allow her one day off a week, which was written into her contract, and then gave her only partial severance pay when she quit, or when Ms. Luo expressed disdain about a Shanghai domestic service company that put nannies literally on display for potential employers to select, as if the women were a new appliance. "It's a violation of our human rights," Ms. Luo said.[1] Eventually the conversation turned to a domestic worker in Hangzhou who three months earlier had made national news and was a top trending topic on Weibo after she had deliberately set fire to her employer's apartment, killing three children and their mother. The nanny, who had acquired a gambling debt, had started the fire hoping to then "rescue" the family and be rewarded financially for her good deed. Because the family was very wealthy—the husband/father, a successful businessman, had been out of town at the time—and the nanny came from a poor village in Guangdong province, this tragedy set off national debates about both social inequality and a lack of regulation of the domestic service industry (H. Zhang 2017; Koetse 2017).[2]

What was of most concern to the women around me that day, however, was how in these debates that erupted on social media, the terrible deeds of one domestic worker were used as "evidence" of the low "quality" (*suzhi* 素质) and lack of morals of poor and rural people (two segments of society that often intersect), even though domestic workers frequently experience forms of discrimination or exploitation that illustrate the exact *opposite*—that is, urban residents' lack of morals and empathy. Yet, domestic workers' daily hardships and the physical and caring labor they perform are rarely considered newsworthy. In the words of Ms. Song, who was also at the NGO gathering that day, "I don't think news outlets care about domestic workers unless it's an incident like this [the Hangzhou nanny]."[3] Their labor also often goes unacknowledged by their employers although "sensitivity to the plight of the Other" is foundational in traditional Chinese ethics (Sundararajan 2014, 184). Indeed, just six months before the Hangzhou incident, Ms. Zhang, a domestic worker from Anhui province, had posted a graphic on WeChat that encouraged urban dwellers to consider and be thankful for all the invisible, taken-for-granted labor their domestic workers do.[4]

Domestic workers' absence from China's news media, aside from sensationalized tragedies, and their limited representation in popular culture (Sun 2009; Kong 2014a), allow the longstanding discursive construction of domestic workers as backward, unreliable, untrustworthy, and uneducated to remain firmly engrained in popular consciousness.[5] In the urban home, as commodified labor and subordinate "outsiders," they are also often denied a voice even as they bear a heavy burden to "prove" their worthiness, not only economically but also, and especially, morally. Domestic worker NGOs for years have organized activities to aid domestic workers and facilitate community and mutual encouragement (*huxiang guli* 互相鼓励) among them (see figure 3.1). They have sought to help them express themselves through singing and writing and have also held public performances, similar to *The World's Factory* (discussed in the book's introduction), which include short vignettes that depict the hardships of domestic workers' lives. In this way, NGOs hope to give them (and other migrant workers) a voice and generate empathy in urban residents (Sun 2014; Wallis 2018; Yin 2020). However, there has been less focus on how domestic workers support each other and represent themselves on social media, as Ms. Zhang did through her WeChat post, even as the internet and social media have been viewed as a means of empowerment for marginalized groups in and outside of China (Han 2013; Shirazi 2012; Wallis 2018; Wang and Liu 2019).

Figure 3.1. Domestic workers at an NGO activity share a laugh at a funny video. Photograph by the author.

In this chapter, I explore how "older" (aged 35 and above) domestic workers, all women and nearly all of whom live in their employers' homes in Beijing, use social media as a space for voice, empowerment, agency, and community. Like the other groups in this book, domestic workers exist in a neo/non-liberal context characterized by the economization of social life, patriarchal gender relations, and an authoritarian state that mobilizes affect for governance and control. Compared to the others, however, their autonomy is often more severely constrained by unequal power relations and their position as subordinated, gendered labor in urban homes, where they are supposed to demonstrate their "scientific" techniques of care, cleaning, and cooking as well as their moral uprightness and dedication to the family's "harmonious" environment despite the fact that they are often the target of suspicion, micro-management, or worse.[6] Given that they must constantly expend much emotional and affective labor in their jobs, what are the affective dimensions of domestic workers' social media use? How do they represent themselves on social media, and how does this presentation contrast with the discursive construction of them in the public sphere? In what ways does social media offer them a space for communicative empowerment, or the ability to have a voice through multiple modes and mediums of expression so that the speaker gains individual and social power (Luthra 2003)? Based on ethnographic fieldwork, involving participant observation, interviews, and casual conversations as well as analysis of their use of Qzone (QQ 空间 *kongjian*) and WeChat, I show that ultimately domestic workers use social media as a space for agency to construct themselves as ethical, empathic subjects who demonstrate care and feeling for each other and those in their small social circles.[7] Because many of them only have one day off a week, the online realm has also become a crucial site for the women to offer each other mutual support and form community. In my analysis, I engage in particular with what Sara Ahmed (2014, 10, 11) calls the "sociality of emotions," where "objects of emotion" circulate, accumulating affective value, in the process transforming "others into objects of feeling."

This chapter continues the book's exploration of the entanglement of social media, affect, and ordinary ethics, or how ethical practices are embedded in everyday life (Lambek 2010, 2015a, 2017), in neo/non-liberal China. After first providing background on the growth of domestic work and the domestic worker industry, I highlight how women, who call each other "sisters" even though they are not necessarily friends, use social media as a site for caring, sharing, learning, and demonstrating their self-growth and forward movement in contrast to public discourses about their backwardness.[8] I also show how, through airing short-lived expressions of grievance, which I call "ephemeral protest," they follow norms of Chinese relational ethics. I argue that, in contrast to the commodified expenditure of affect and emotion in their jobs, this self-representation and expression is empowering and allows them some degree of agency. I say "some degree" because, while they do engage the government's top-down discourses of positivity and harmony, and any affective labor entails costs, they use social media as a space to build community, generate hope, and express pain and recovery, thereby producing affective value for themselves and others. In this way, social media is articulated to their communicative empowerment, but this too is only partial because, while the women value each other's voices, their social media practices do not necessarily lead to "voice as value," or their voices being valued in the larger society (Couldry 2010).

Domestic Workers, Commodification, and Exploitation

Domestic work emerged at the beginning of China's reform period in the early eighties as an occupation deemed suitable for rural women and urban women workers who had been laid off from state-owned enterprises that closed as a result of China's market reforms.[9] Domestic work was viewed by the government as a way for these women, who had a lower education level and a skillset viewed as incompatible with the new market economy, to escape poverty. For rural women in particular, it was seen as a way for them to improve their *suzhi*, or their cultural, psychological, and moral "quality." As Yan Hairong (2008, 114) argued, at the time, *suzhi* was coded as a "a new form of value that represents human value to the teleology of development."[10] This process of coding human value and its articulation to *suzhi* still persists.

In an interview, Ms. Xu, a domestic worker originally from rural Hebei, explained her reason for being in Beijing for more than seven years:

> It's hard for my *laoxiang* (老乡) [meaning people from the same area] to stay here. There are so many problems, and they can't resolve them. But we must earn money so that our kids can go to school. Still, there is an idiom, "Staying in your hometown is always better than going out." But our families need the income. . . . Young people think it's an opportunity, but for us, we don't have other options.[11]

Unlike younger migrant workers who go to cities to gain new experiences and some autonomy (Gaetano 2004, 2015; Wallis 2013a; Wang 2016), Ms. Hu and women like her, who are usually in their mid thirties to fifties, are often working to support children or elderly parents back home, much like female domestic workers in transnational contexts (Hondagneu-Sotelo 2001; Parreñas 2008). These women's labor is a sacrifice for their families, yet most, especially those from rural areas, view their experience as an opportunity for self-development. Just as in the past, state discourse, often disseminated through domestic service agencies, encourages this view, framing such labor as an opportunity for these women to transform their subjectivity and improve their *suzhi*. NGOs that offer various kinds of training (English classes, labor rights workshops, creative expression) aid the women while also focusing on such "self-development." They also try to give them a voice, yet one that is often constructed by the NGOs (Jacka 2006; Sun 2014; Wallis 2018; Yin 2016).[12]

Early in the reform period, urban families who were becoming part of the emerging middle class employed domestic workers as a means of demonstrating their high *suzhi* and class distinction (Sun 2009; Yan 2008). Over time, having a domestic worker in the home became seen as commonplace by members of the urban middle class. The exact number of domestic workers in China is difficult to determine given the informal nature of most domestic work, but in 2021 it was estimated at 37.6 million (iiMedia 2022). Half of these workers reside full-time in their employer's home.

Emblematic of the increasing commodification of all aspects of life in neo/non-liberal China, domestic work has become a booming industry with more than 2.65 million domestic service agencies (Mo et al. 2022).[13] Numerous private companies—some quite small, some large—have emerged to take advantage of this economic opportunity (see figure 3.2). As a result, the government has tried to standardize the industry, which has had several consequences for women in this occupation. With standardization has come a number of specializations, including maternal care, elder care, and infant care.[14] This segmentation has increased potential employment opportunities, but it has also given rise to professional standards that require different types of instruction and certifications that women must obtain to make themselves marketable (Fu, Su, and Ni 2018; Tong 2018). Training is meant to impart certain skills based on "scientific" methods that potential domestic workers apparently cannot learn on their own; it also disciplines the workers through instructing them to have a positive attitude and maintain harmony in the home regardless of how they are treated (Fu, Su, and Ni 2018). Moreover, in news media and surveys, domestic workers are frequently criticized for lacking education, which reinforces the perception that there is a need for such training (see iiMedia 2022; "Jiazheng Fuwuye" 2018). Finally, the rise of domestic service agencies

Figure 3.2. A domestic service agency in Beijing. Photograph by the author.

and the division of domestic work into these different categories has actually led to a decrease in salary for women employed in certain segments, such as maternal care, because their individual bargaining power has been reduced (Qian 2022).

Like most industries in China, the domestic service industry has become integrated into the digital economy. Especially since the implementation of China's Internet Plus initiative in 2015, the government has worked with domestic service companies to help in their digital transition (iiMedia 2022).[15] Potential employers' use of online domestic service agencies and social media apps has increased because it is seen as convenient and more reliable. Domestic workers also use these apps and other websites to find information and search for employment. However, with more companies using social media platforms to promote services (training, certification, job placement), more opportunities have emerged for unscrupulous companies to take advantage especially of rural women, who have minimal computer literacy and little experience in negotiating contracts.

As just one example of how platformization can harm domestic workers, while I attended an NGO workshop on domestic workers' legal rights, a woman shared a story about paying 1,800 *yuan* (approximately US $282 at the time) to participate in a training course in infant care offered by a company she had found online. At the end of the training, she received a certificate but was told she would not be placed in a job until she completed another class for an additional 1,800 *yuan*. "The training was useless," she said, even though the company was verified as legitimate on a government website.[16] Other domestic workers told of signing vague contracts with placement agencies only to find that once they had started the job, the requirements differed from those to which they had agreed. However, in such situations domestic workers have little legal recourse.[17] The state has created policies meant to regulate and grow the industry, yet most domestic work is still not covered by China's Labor Contract Law because it is considered informal employment (Hu 2011; Ma 2010; Qian 2019).[18] When a scandal or a tragedy arises, such as the fire in Hangzhou, it is used as a reason to implement more controls, yet these tend to protect agencies and urban employers, not the workers themselves (Fu, Su, and Ni 2018; Tong 2018).

Relational Norms, "Cold Violence," and Gender Hierarchies

It is not just state policies and corrupt agencies that can disadvantage women who seek employment doing domestic work. Live-in domestic workers are easy targets of surveillance and rarely have time to themselves (Ellerman 2017; Gaetano 2004; Yan 2008). Aside from physical labor, their lives are characterized by emotional labor, Arlie Hochschild's (2003) term for the particular type of commodified labor that service workers (especially women) in the public sphere are expected to perform, such as making customers feel good through smiling and maintaining a polite demeanor while masking tiredness or frustration.[19] They also expend affective labor, or labor involving care, human contact, and the "creation and manipulation of affects" and which is meant to produce "a feeling of ease, well-being, excitement, passion—even a sense of connectedness or community" (Hardt 1999, 96). These terms, although analytically distinct, highlight how care work commodifies affect, emotion, and intimacy.[20]

Chinese live-in domestic workers have experiences similar to those in other settings (see Hondagneu-Sotelo 2001; Huang and Yeoh 2007), yet they are situated in a particular socio-cultural context undergirded by historical relational patterns regarding who is considered an insider or outsider. As outsiders with low status, the women are often not thought to deserve the same treatment as those in their employers' immediate social networks. As elaborated upon in my discussion of the rural micro-entrepreneurs in the previous chapter, family and kinship ties determine rules of attachment and obligation; hence, it is difficult for most domestic workers ever to feel part of their urban employer's family regardless of how long they have been there, and even when the family tries to treat them as a family member (Gaetano 2004; Sun 2009). Sun Wanning (2009) has elaborated on domestic workers' "ubiquitous invisibility" and their position as "intimate strangers" in the urban family home, where there is often a large cultural and emotional distance between them and their employers. This situation is compounded by the fact that the women's labor is seen as a market transaction. As one NGO staff member said in an interview:

In traditional Chinese thinking, domestic work is not considered "real" work. . . . people think of them as servants with the lowest social status.

Even now this traditional thinking continues. You can contrast this with other countries, where domestic workers are seen as helping the family and the family thanks them. A mother is thankful to someone who takes care of her baby. But in China, the thinking is, "If I pay you money to take care of my baby, you should listen to every word I say, and you are lower than me." It's like being a slave.[21]

Domestic workers I knew shared experiences that exemplify this idea of "being like a slave." Ms. Luo stated in an interview, "My life is basically to do whatever my employer wants me to do."[22] Another woman said that she felt like she should do extra work all the time, regardless of the terms of her contract, because she lived in her employer's home. Other domestic workers expressed indignation when they were only given leftovers to eat, a common complaint that has generated public debate in China over the years. These examples reveal how, as opposed to physical violence, domestic workers must contend with what this same NGO staff called "cold violence," meaning emotional or symbolic violence in the family.[23] In such situations, defined by commodified labor relations, domestic workers are nonetheless expected to regulate any negative emotions as they perform physical and emotional labor.

The boundaries between domestic workers and their employers due to the workers' outsider status are reinforced by the intersection of class and rural/urban biases and patriarchal gender hierarchies, which especially stifle the voices of rural women who are domestic workers. Ideas regarding women's subordinate status, along with the notion that their duty is to be "virtuous wives and good mothers," have been recuperated in official and popular discourse in the reform era (they never disappeared in rural areas), and increasingly so under Xi Jinping, as discussed in the book's introduction. Domestic service agencies, both private and state-run, draw on such discourses in their training, which ultimately function as modes of governance and control. In particular, rural women, who have been subordinated in their families in their villages and who are discursively constructed as not knowing how to speak, are then rendered powerless in the urban home. Domestic workers' voicelessness and powerless are compounded by Chinese norms of how to complain, which dictate that unless a grievance is with a close friend or family member, it is more appropriate to deal with a complaint indirectly (Gao and Xiao 2002).

During workshops and other NGO-sponsored activities I attended, domestic workers who told stories of being constantly micro-managed or the subject of accusations that made them lose face felt like they had few means to talk back. As just one example, Ms. Li had quit a previous job because the father of the household noticed a small bruise on his young son's leg after the boy had been in her care. She said, "He yelled, 'Look what you've done! You are terrible at your job!' But I did not injure the kid. It's common for a child to get a bruise when they're playing." She then added, "I quit after that. I would never work for that family again."[24] Like many domestic workers who feel unfairly targeted, she did not contest the accusations. Her silence was both an affective and ethical response: it was born of anger and shame, a deep sense of injustice, and a recognition of her powerlessness. Quitting was then her only recourse.

As this discussion makes clear, the situation of China's domestic workers exemplifies Hemmings' (2005, 561) argument that certain marginalized subjects are "so over-associated with [negative] affect that they themselves are the object of affective transfer." For domestic workers, this negative affective transfer marks them as "other" and thus not worthy of mutual care and empathy. Even when they are treated well, there is constant pressure to perform the role of devoted, self-sacrificing caretaker. In the remainder of this chapter, I discuss how they use social media to affirm their moral character and "talk back," usually indirectly and in a variety of ways, to such negative discursive constructions.

Objects of Emotion and Affective Economies of Care

Despite the struggles domestic workers face, on social media much of the women's self-expression emphasized positive, rather than negative, feelings and experiences. In one sense, such "performative positivity" (Wallis 2022) could be understood as a manifestation of the psychologization of Chinese society, which emphasizes happiness and self-growth (Yang 2015; Zhang 2015, 2020); it also seems to conform to the government's desire that "positive energy" be the dominant sentiment online. However, rather than reading this stress on the positive as evidence that the women blindly followed state discourse (in contrast to the oppositional stance of the young feminists featured in chapter 4), a deeper analysis reveals that their purpose was in part to counter state and popular discourses regarding their "backwardness" and "low morals" by demonstrating themselves as ethical, caring, and empathetic subjects.

Often in studies of social media platforms, the large degree of affective (immaterial) labor users expend is understood as ultimately for the sake of capital accumulation by tech companies and secondarily as expanding the cultural capital and building the "brand" of the user (Coté and Pybus 2007; Terranova 2000). Of course, these factors exist in China, especially given the enormous sums of money social media influencers make for themselves and the platforms (Powell 2021).[25] I argue, however, for a different understanding of the women's social media use. Among their small networks, their social media content, as expressions of voice and as Ahmed's (2014) "objects of emotion," circulated and generated feelings and attachments; it stirred *gan* (感), enabling them to feel helpful, appreciated, and part of a supportive community. In the process, these affective economies produced value for themselves and in relation to others.[26] On QQ and then later WeChat, affect and emotion flowed as the women constructed their profiles, performed virtual care work, displayed skills and learning, and occasionally voiced complaints through what I call ephemeral protest, or a grievance that evaporated, in the sense of not being dwelled upon, after it was recognized.[27]

Profiles

When setting up a Qzone or WeChat profile, users create a name, and they also have the option of composing a signature (or motto). The domestic workers chose names like "Happy Dove," "Better and Better," and "From Happy Sichuan," and mottos like "Laugh at Life and Tread Lightly," "Forget Troubles, Be Happy Everyday!" and "My Life I Call the Shots!" Such names and mottos emphasized positive feelings and social media as a space for agency, or "calling the shots," in contrast to their place in their employers' homes. Some women also opted to write an explicit moral message. For example, Ms. Dong had recently quit a job as a domestic worker, and like Ms. Li, was not given all the pay she was owed. For her motto she wrote, "People work hard to make money their whole life, [doing] all kinds of jobs, all for one thing, and that thing is money. Money is evil." In distancing herself from such "evil," she implicitly called out her employers' character (or lack of). The women's profile images were also carefully chosen. These were rarely of the women themselves except in a few instances. Instead, the women chose cute avatars, flowers, or something else considered "fun or beautiful."

As these mottos and images could be the first information seen by someone browsing the women's profiles, they were intended to give an impression of an agentic, morally upright person. In a sense, the women's profiles demonstrated the "impression management" that is frequently seen on social media. However, my concern is how the profiles were one articulation of the positive affective value that circulated among a small number of contacts (20 to 30), some who were emotionally close, and others who were bound together through shared circumstances. These included family members, migrant workers—especially domestic workers—and NGO staff, but rarely employers. Overall, the women's profiles revealed a desire to express outwardly an optimistic and even carefree outlook, in clear contrast to the reality of their lives.

Virtual Care Work

Domestic workers' lives are largely defined by gendered, caring labor performed for both their employers and their own families back home. Regarding the latter, much scholarship has shown how domestic workers in China and in transnational contexts have used communication technologies to attend to their family members, especially children, at a distance through calling and text messaging (Cao 2009; Parreñas 2005; Thomas and Sun 2011; Uy-Tioco 2007).[28] It has also revealed how smartphones allow for "intensive mothering"—video calls, chat messages throughout the day and night, and surveilling children through using social media platforms (Madianou 2012; Chib et al. 2014; Platt et al. 2016). The goal of most of this research has been to understand relatively *private* interactions between domestic workers and geographically distant family members. In contrast, elsewhere I have discussed some of the very *public* ways that Chinese domestic workers use communication technologies to engage in performative motherhood (Wallis 2022).

Here I focus on the women's caring labor that was directed outward so that all in their social media circles would benefit, in particular through pedagogical and uplifting posts. Most of these posts contained forwarded content that was deemed "useful . . . informative," in Ms. Zhang's words.[29] Several women also said they emphasized positivity in their posts because "no one wants to be surrounded by negative things." Again, this focus should not be read as merely complying with the party-state's desire for "positive energy," which is a regulative discourse in neo/non-liberal China. Rather, among women whose daily lives were often difficult and challenging, social media could provide an escape, not in the sense of fantasy, but by being a realm for sharing, caring, and encouragement as an ordinary ethical, other-centered practice.

Pedagogical posts meant to show the domestic worker as a nurturing and discerning caregiver, which were the most common, continued the women's affective labor into the online realm. Reproductive labor manifested in tips about foods that were good for children and advice on handling an infant. There were also topics related to hygiene, such as guidelines on how to properly wash one's hands or handle raw eggs. Various remedies, such as the use of traditional Chinese medicine for different ailments, tips for staving off illness, and exercises for poor blood circulation, were also common among several women, especially those in their forties or fifties.[30] Occasionally, these were accompanied by a comment, such as when Ms. Wang posted an article about cancer prevention and wrote, "We are at the age where we need to pay attention." Through such "how to" content, domestic workers demonstrated their other-centered ethic of care and concern. In turn, the value of such posts was affirmed through "likes" and sometimes brief comments such as "Useful information" or "Important to know" by other domestic workers.

Closely connected to pedagogical posts that embodied reproductive labor were posts that highlighted self-care, both aesthetically of one's body and physical appearance (proper skin care, hair styling tips) and spirit (handling heartbreak with dignity, the qualities of a good woman). For example, one forwarded post advised, "If a man neglects you leave him. Take care of your skin. Dress elegantly. Don't drink or smoke. Vanity is poison. . . ." Another item, which several women circulated, was titled "Xi Jinping's Letter to All Women." In this letter (possibly a hoax), China's president extolled the virtues of tending to one's partner's spirit rather than to material goods or physical appearance. In these examples, much of the advice was infused with notions of binary, essentialized gender. That domestic workers posted Xi Jinping's words as those they should follow could certainly be understood as demonstrating the effectiveness of the states' mobilization of affect for social control and stability. In other words, the posts implied not just agreement with the sentiment but also approval of the messenger. Another interpretation, however, is that the posts were performative, affirming a gendered identity in which the individual is deeply invested and for which they are rewarded. Of course, both, and other, interpretations can exist simultaneously.

In contrast, other posts that focused on self-care revealed the widely diffused psychologization of Chinese society and the role of indigenous philosophy and religion in this process. These ideas are deployed as a mode of governance among elite members of society, and they have been taken up by middle-class urban Chinese individuals to deal with feelings of purposelessness and unhappiness (Yi 2019; Zhang 2020; Y. Zhang 2018).[31] The rapid manner in which religion was revived in the countryside at the beginning of the reform era, after being quashed by Mao Zedong for thirty years, also reveals the importance of such practices among diverse members of society (Yang 2011). Indeed, people seek out beliefs and rituals that help them deal with the stress and turbulence that are constitutive of China's ongoing transformations. Over the course of this research, three domestic workers I knew became interested in Buddhism and occasionally posted articles or graphics with Buddhist content. One woman wanted to share information about Buddhism and ways to cultivate mindfulness because she found it helpful in her life in Beijing. Domestic workers, as extremely marginalized, are often imagined by the urban middle class as only focusing on economic survival, yet they and many other migrant workers also desire more meaning and spiritual purpose, as many NGOs have recognized (Yin 2020).

Finally, caring labor also manifested in posts filled with what I call "words of wisdom."[32] Such posts contained moralizing and sentimental messages that highlighted qualities deemed necessary for survival in neo/non-liberal China's hyper-competitive world of marketization and self-responsibilization. Such qualities included having self-confidence and resilience while also maintaining the ability to be happy with one's lot in life. For example, Ms. Qin advised, "Happiness can be contagious.

If melancholy is like an occasional dark cloud, then happiness should be regarded as occasional sunshine that penetrates the cloud." Trusting in fate and having faith in the future, particularly when facing loneliness or alienation, were also crucial, as in Ms. Sun's forwarded post: "When I'm tired and lonely and I want to give up, and there are too many sad things . . . and too many feelings of helplessness, life still continues. The future is still waving to me. Come on. . . . Continue to move forward! Never look back!" Real friendships, characterized by true feeling (*ganqing* 感情) and the ability to offer support in times of hardship, was also a common theme. For example:

> You don't need to have many friends as long as their hearts are sincere. . . . Whether in daily life or online, we have to have friends. In life you'll encounter difficulties, so sincere friends are more valuable than gold; sincere friendships are priceless. When you are hurt, your friends will give you energy and stay with you until you recover. . . .

Like Hallmark greeting cards, such posts were characterized by sentimentality and optimism meant to mobilize positive affect even when one might feel lonely, tired, and hopeless. These messages were not just for the women themselves, but also, several women said, to provide comfort to others, again revealing their empathy and other-centered ethics of care and concern for those in their community.

Significantly, most of these "words of wisdom" posts are written by employees of state-owned telecommunications companies or corporations toeing the Party line. Their melodramatic flavor is meant to cater to a social media user with a low-income level and who is older and less educated (Qiu 2009; Wallis 2013a), and they exemplify how affect and governance meet. Their writers also engage in affective labor, or the manipulation of affect, emphasizing hope, positivity in the face of adversity, and individual resilience rather than highlighting structural factors that cause personal difficulties. Many of the posts about self-care also had this purpose. As they circulate, they accumulate affective value that in many ways supports the neo/non-liberal state and the women's own marginalization. However, in my fieldwork it was clear that they were objects of feeling and had value for the women because, through such posts, they demonstrated their care and support for others in circumstances that they as individuals felt they had no power to change.

Displays of Learning and New Experiences

During a fall day in Beijing, I noticed that many women had shared a post (and one woman had shared it with me) about Peng Liyuan, Xi Jinping's wife, in which she was praised for having reached her current position not because of her husband but because of the "three most valuable resources she cultivated in herself": maintaining her physical appearance, continuously learning and growing, and working hard. Reading further, one is told that continuing to learn and grow is considered important for setting an example for one's children and becoming "the forever muse of men." Peng Liyuan has long been considered a role model in the public sphere, and her physical appearance, including her attire, hairstyle, and mannerisms, have drawn much attention because they contrast with the bland appearance of the wives of former Chinese heads of state. Once again, the essentialized gender present in the article is not surprising, particularly given that Peng's public persona, including her statements on marriage, exemplify such essentialism. The message of continuously learning, growing, and working hard also reflects the self-responsibilization necessary to succeed in neo/non-liberal China. It connects as well to a second major theme found in the women's posts, which is how domestic workers use social media to display their own self-development.

In neo/non-liberal China and elsewhere, the economization of social life requires that individuals constantly make themselves more valuable and marketable, which is achieved by attaining new skills, certifications, etc., as I discussed above. In interviews, several domestic workers spoke of their lives in Beijing as difficult yet nonetheless an opportunity to learn and develop. In contrast to young migrant women working in service work or factories, who tend to use mobile and social media to demonstrate the transformation of their outward appearance (Wallis 2013a; Wang 2016), these older women were more likely to display inward transformations in the form of newly acquired skills and talents, continuous education, and fresh experiences. Like the caring labor just discussed, these posts were infused with positive affect. Given the heavy burden the women bore to "upgrade" themselves, such displays were performative; that is, through demonstrating they had worked on themselves in the city, they called into being the women's "quality," thereby countering the discourse of their "backwardness." This sharing of skills was also another manifestation of the sociality of emotion and of domestic workers' maintenance of community and support of one another.

Many posts in this category focused on food, which is important in all cultures. However, in China food has special significance, connected as it is to specific regions, histories (including severe famines), wealth, and status. Cooking and feeding others is, of course, also a central part of a domestic worker's job, so in some ways it is not surprising that several women posted images of dishes they had cooked. This practice is also widely found among a variety of demographics in and outside China, as anyone who uses Instagram is fully aware.[33] Although the domestic workers wanted their friends and family to see dishes they had cooked and of which they were proud, they also had other motivations. Talking about how she decided what to post on WeChat, Ms. Zhao said, "I post different content to different groups. In the domestic workers circle (*jiazhengquan* 家政圈), I post how to make good food and other skills so that we can learn from each other. I also forward articles about cooking and how to make good dishes."[34] Ms. Xie echoed this sentiment in an interview:

> [On WeChat] I post delicious dishes, for example *baozi* (steamed dumplings 包子) and Beijing crispy fried pork. I make the dishes and then if they look good, I post pictures to show my hometown people (*laojia* 老家). I learned how to make [the dishes] from a WeChat video and then I showed them [so they could see how to do it]. I made it myself, and I thought, "Very good."[35]

In chapter 1, I discussed Bo, a young creative who expressed disdain about his rural peers back in the village, whose "posts on Weibo or WeChat are just about the food they cooked and/or ate, their family, or cute animals." Bo saw these posts as evidence of his peers' stasis and lack of ambition. Here I argue that domestic workers' posts of food showed the exact opposite; that is, forward movement and learning, particularly since the dishes Ms. Xie mentioned are northern cuisine and not something she would necessarily have known how to cook prior to living in Beijing. These dishes were prepared for her urban employer, and she was clearly proud of her success and enjoyed the good feeling it brought her. Similarly, Ms. Zhao's other-centered focus meant that she also wanted other domestic workers to learn because it might benefit them in their jobs (if they cooked such dishes and their employers appreciated it).

In addition to learning and displaying recipes and food, a continuous commitment to other types of informal education, which were not connected to domestic work, was also evident. For example, Ms. Zhang, who said her posts had to be informative and helpful, frequently shared information on how to gain new skills, such as improving one's handwriting through practicing a word a day (see figure 3.3). This type of self-enhancement differed from the modes of self-care discussed above because the focus was on tangible skills that might increase one's cultural capital and possibly make one more marketable. Some posts were explicitly about learning how to do something, such as direct marketing, that would generate income. Only a few women posted this type of content, however. When I asked one woman if she had tried direct marketing, she said she had not but merely wanted to circulate the information.[36] On occasion, learning skills was connected to hard news, such as when some women shared a quasi-news/opinion piece on the five types of jobs that will be eliminated in the future in China. The lingering reminder of the need to constantly upgrade oneself was tangible in the post. When I asked Ms. Li about her motivation for posting it, she said, "It's important to be aware."

In addition to new skills and practices, new experiences were also shared, often in a positive and uplifting manner where the women's own precarity was forgotten, at least momentarily. For example, during a trip with her employer's family, Ms. Xu wrote, "Today we ate in an international hotel. When I was small, I wanted to see the ocean, and today I finally saw it! The rippling of the endless blue sea made me feel happy. All wishes come true!" Such experiences were usually documented through taking pictures. If a woman had a chance to go to a park or other scenic or historical attraction in Beijing, the photos were posted on social media.[37] These posts stood out because they were relatively rare for most women. Often the activity was sponsored by an NGO, and thus there

Figure 3.3. A screenshot of an advertisement for a program to improve handwriting. Screenshot by the author.

were pictures of women in groups, and these types of photos far outnumbered pictures of family members, revealing how the NGOs provide a sense of community far from home.

Without exception, photographs of new experiences focused on leisure, pleasure, and relaxation. Often the images were accompanied by short comments, such as "Look at this beautiful scenery . . . [as we] all jumped together" or "A New Year's Day get-together. Good fun and a learning experience. I am thankful for everyone in my life." Many photos were of two or three women together. Only a few women posted pictures of themselves, yet these were rarely selfies (unlike the abundance of selfies snapped by younger migrant workers).[38] The photos were always taken by another person and often grouped together in a series that had some pictures of the woman alone and others of her with other "sisters." The only exception was Ms. Li, who enjoyed singing traditional opera from her home region. One day, after she had quit her job because her employer refused to give her a day off, she decided to don some colorful Chinese clothing and go to Tiananmen Square, "for fun," she told me. She was thrilled when a group of international students there wanted to take pictures with her. She then posted several of these images, with and without the students, on her WeChat Moments. Did the students exoticize her and/or did Ms. Li see her value in the fact that they wanted to take pictures with her? Perhaps, but another interpretation is that the good feeling that was produced generated affective value for all involved and left an "impression" (Ahmed 2014) that potentially created affinity by breaking down boundaries between people from very different backgrounds.

Through sharing pictures of food or newly acquired skills or images of special experiences and opportunities, the women demonstrated processes of self-transformation and the construction of the self in a particular way, which in and of itself was an empowering practice. Although these were individual expressions, they were highly relational, and the images, as objects of emotion, especially required a response—a "like," a thumbs up emoticon, and/or brief comments of praise or encouragement from others in their network, revealing how the sociality of emotion requires participation for affect to be effective.

Ephemeral Pain and Protest

The social media uses discussed so far overwhelmingly emphasized positive affect and emotion. These posts generated a mixture of *ganqing*—positive feelings for the woman who posted—and *youqing*—in relation to others (Y. Zhang 2007). Still, these could be viewed as

exemplifying how the party-state in neo/non-liberal China has effectively harnessed affect for purposes of control and value extraction. In
other words, women who regularly face challenges, and in some cases
extreme exploitation and discrimination, have been pacified through
discourses of "positive energy" and "harmony." However, this view
neglects to consider the structural factors that shape older migrant
women's experiences and their relative powerlessness. It also ignores
how Chinese people tend to express grievances or complaints, which
depends on the degree of closeness between parties and their relational
position (Gao and Xiao 2002). Only in very close (familial) relationships
should one state a complaint directly.

Below, I discuss how some of the domestic workers did indeed voice
their frustrations on WeChat, but first it is helpful to situate such expressions in their offline practices. An edited excerpt from my fieldnotes
written during a Saturday NGO meeting provides one example:

> At the start of the morning workshop, a group of five new women arrived.
> At first, no one was sure who they were because the domestic worker
> who had invited them did not arrive until later. Nonetheless everyone
> welcomed them, and the workshop got started. . . . During the break after
> lunch, we all sat in a circle and one of the new women complained at
> length about her employer. At first people offered words of support, but I
> noticed that eventually several of the regular members stopped engaging
> in the conversation and in fact looked bored. Just before the afternoon
> activities were to begin, the five new women and the woman who had in-
> vited them left. After they were out of earshot, one of the regulars turned
> to the others and said, "*Tai chao, a?*" to which some silently nodded their
> heads in agreement.[39]

As mentioned earlier, during NGO-organized activities, I frequently
heard domestic workers vent about their situations, but such complaints
were never expressed to the degree that I heard that Saturday morning.
The reason is that doing so violates a key aspect of interpersonal ethics, which is care and empathy for others. The meaning of "*tai chao*" is
"too loud," and in this case, so talkative and negative that it creates an
unpleasant atmosphere for those around the person who is complaining.
As equals in status, living in similar circumstances, the women could

openly share negative feelings and emotions with each other, but unspoken norms required that others' feelings also be considered in the process. Domestic workers viewed the Saturday workshops as a time to unwind, relax, and have fun. Incessant complaining, with no purpose other than for the speaker to vent, demonstrated the speaker's lack of concern for others. These norms of direct/indirect speech, as they are informed by relational closeness and power hierarchies, also explain why domestic workers usually do not express their grievances directly to their employers and opt for indirect means, such as quitting a job when they feel humiliated or cheated, rather than reporting an exploitative agency or employer, as mentioned earlier.

Domestic workers used social media to express negative feelings when the hard reality of their lives became overwhelming, yet these posts were much smaller in number than the other types of posts. At varying times, all the women voiced feelings of sadness or loneliness. For example, Ms. Luo wrote of being sick with no one to take care of her, to which other domestic workers responded with short, empathetic comments. After Ms. Sun described how she felt sad because she could not return home for Spring Festival, another domestic worker wrote, "Little Sister, Happy New Year! Turn your grief into strength. I understand your feelings, (we must) be strong for our children. Keep on! Life will be even better later!" Sentiments such as these almost always generated replies, which offered encouragement and reiterated the need to be strong and persevere. The responses tended to focus on uplift and resilience, not victimhood or pity.

In addition to voicing feelings of pain from being alone, a few women also shared with their broader WeChat contacts the frustration and humiliation that could accompany domestic work. One woman who posted frequently about being misunderstood by her employer, one day wrote, "I truly know they won't trust me. Why should I be this angry?" In another post she stated:

> Domestic workers are very tired [and] busy all the time. When employers are happy, you need to smile. When your employer is angry with you, you have to keep silent and tolerate it. Who can understand domestic workers? My life is so difficult, but who cares? If I had a choice, I wouldn't be a domestic worker, but who could my family depend on if I didn't do this

work? So, no matter how difficult or tiring, I'll persevere for the sake of the elderly and the young in my family. I have to face all the injustices. One day I'll be finished and have my own life and have a better life than others.

With her words, she emphasized her moral obligation to her family. She also eloquently captured key aspects of Hochschild's (2003) definition of emotional labor, such as masking one's emotions for the sake of others' well-being, and the emotional exhaustion it brought. As Hochschild theorized, over time, as outward display does not always match inward feeling, "emotive dissonance" can lead to stress and self-alienation (90). This post, circulating negative affect, generated empathetic comments from other domestic workers, including one who wrote, "I see this and my heart aches." Yet in order not to be "*tai chao*" in the figurative sense, and to balance out the negative feelings, a few days later the woman wrote that she was touched that the grandmother in the family had bought her mooncakes during Mid-Autumn Festival, something her employer had never done before.

In contrast to content shared to all of their contacts, some grievances were only voiced among those deemed close enough to bear a burden, meaning circles of close friends or family. I could not view these posts, but in interviews my informants occasionally discussed them.[40] After an NGO English class one Sunday, Ms. Wu told me and the other domestic workers present about her employer, who made her sleep on the balcony of their apartment. The balcony was enclosed, but there was no heat or electricity. During the winter, Ms. Wu's employer only gave her a hot water bottle even though it was quite cold. After Ms. Wu asked for a quilt, her employer reluctantly agreed. "If they aren't using the quilt, they should let me use it since it doesn't cost anything," she said.[41] Despite discussing her situation with us in person, Ms. Wu later told me she only shared this predicament with her hometown group on WeChat. I asked her how they responded, and she replied, "They comforted me. One of them said, 'At least you have a place to stay. . . .' Together we all share our anger, sadness, and joy." What is striking about this comment is how it focused on what one should be grateful for, even in difficult circumstances. It exemplifies the internalization of disciplinary discourses that dictate that domestic workers should suppress any negative emotions, however justified such feelings are.

The experiences of these women and other domestic workers confirmed the importance of social media as a space for them to offer and receive mutual encouragement and support. These women aspired to be treated like members of their employers' families, but this hope was indefinitely deferred. Because employers view them as "other," or in the worst case, "like a slave," they routinely neglect common compassion without shame. Were domestic workers actually treated like family members, they could express their discontent at least somewhat directly.

Employers were not the only ones who violated ethical norms, however. One day, Ms. Chen, a domestic worker from Anhui province whom I had known for a few years, invited me to join her for lunch because she was leaving Beijing the next day to return home. A few days later she wrote about how she had been severely let down by a "close sister" who never showed up when she was supposed to help Ms. Chen carry several pieces of luggage to the train station. Ms. Chen felt particularly betrayed because she had just helped this woman move to another home. Then, when Ms. Chen finally got on the train, people scolded her for having too many bags. She ended her story:

> I wish next time I travel by public transportation, if I have to bring luggage, I'll only bring one thing: renminbi [Chinese yuan]. I hope I don't have to go out and do migrant jobs once I get home. I hope that at home I'll have cars when I go out, have money when I come back, make money when I work, [and] have friends when I hang out.

Ms. Chen's deep sense of feeling let down arose not only because of the breach of trust by a "close sister" who assured her she would help, but also because this sister did not follow the norms of reciprocity expected in Chinese social relations; e.g., Ms. Chen had just helped her, so she was obligated to help Ms. Chen. Although domestic workers form community with one another, often these are tenuous bonds, defined by shared experiences of migration, particular forms of labor, and marginalization. Ms. Chen's anger and sadness was thus compounded by her awareness of her limited ability to cultivate both social and economic capital in the city.

The posts of pain and protest clearly demonstrate the hardships and stress faced by domestic workers. I call them ephemeral, however, because they were almost always tempered by a subsequent post that was happy or uplifting, meant to generate a flow of positive feelings. This tendency mirrored their offline practices as well. Aside from the women who were called "*tai chao*," in group settings when a woman would vent about something rather extensively, she would frequently end by saying something positive. If the complaint could somehow be interpreted as criticizing China, she added a comment about loving China, perhaps due to my presence lest I interpret her words as unpatriotic. Of course, such statements also show the depth of top-down governance, where anything seen as an affront to Chinese people, culture, or the nation is stifled.

Conclusion

Lawrence Grossberg (1992, 83) has argued that "it is in their affective lives that people constantly struggle to care about something, and to find the energy to survive, to find the passion necessary to imagine and enact their own projects and possibilities." In this chapter, I have explored how live-in domestic workers, who are middle-aged women mostly from rural areas, use social media as a space to create community and construct themselves as caring, empathetic, and ethical subjects, in opposition to how they are often represented in official and popular discourse. Their social media use is articulated to their position in the urban home, which is defined by commodified labor relations that often do not produce the norms of reciprocity that should ensure their ethical treatment by their employers. Without this reciprocity, domestic workers are required to make everyday ethical choices, for example, how much physical and affective labor to expend on top of the basic requirements of their job, whether to defend their honor when unjustifiably accused of wrongdoing, and if and how to report a morally suspect agency or employer. In such situations, most domestic workers choose either to stay silent, regulating any negative emotions to maintain "harmony," or to "vote with their feet" and quit. Despite such experiences, on social media the women overwhelmingly chose to emphasize the positive: in their profiles, through virtual care work, and through mutually supporting one another. They also affirmed the role of a domestic worker as someone learning and gaining new skills and experiences. Even their expressions of grievances revealed an ethical commitment to those in their WeChat friends' circles. Social media thus provided them a space for agency and for communicative empowerment. Through their affective labor, affect and emotion circulated, and these affective economies were invaluable in sustaining their mental health and well-being.

In some ways, my analysis of the women's social media use could be seen as too optimistic, and not adequately addressing the hardships of their lives or the affective attraction of the top-down discourses of positive energy and harmony in neo/non-liberal China, where even women who face difficulties on a nearly daily basis nevertheless emphasize care and uplift in their social media use. My focus emerges from their mediated voice and virtual self-presentation as well as from my interactions with numerous women. Yet, it is also a strategic choice. The women certainly recognized their own precarity caused by structural inequalities far beyond the level of the individual urban household, yet their particular exertion of affective labor via social media can be read as a strategy of counter-representation and a moral commitment to themselves and each other.

Michael Hardt (1999) argues that affective labor, because it promotes a sense of community, embodies the potential for social change. However, Kathleen Stewart reminds us that "all agency is frustrated and unstable and attracted to the potential in things" (Stewart 2007, 86). Within relations defined by inequitable power structures, the women used social media as a space for sociality, belonging, and support, as affect motivated passions and produced alliances between women from different geographical regions and backgrounds, which was reinforced and grew online. In this way, the potentiality of affect is apparent in their social media use.

The women's emphasis on positivity is not, as some have argued in the context of western, middle-class social media users, due to an inordinate concern with managing a self that is outwardly fulfilled and happy. Rather, it was clear in interviews and conversations that the women understood their social media posts as a mode of caring for those in their social networks. The importance of such crucial social support should not be minimized. As one woman said, "In real life everyone is so busy, so our social life is on WeChat."[42] Still, this calming and caring function could be seen as muting any form of resistance. However, this type of reading implies the women are duped and cannot see their own subordination, which would not only be an elitist, condescending understanding of their social media practices and strategies but also would deny their agency. It would understand agency and resistance as zero-sum games. Still, the individualized communicative empowerment I discussed in this chapter *is* limited because there is no way to claim that it is socially transformative; that is, it does not lead to voice as value and structural changes that would shift unequal power relations and improve domestic workers' lives. In an online atmosphere that is increasingly censored and surveilled, the dominant concerns they express are not politically risky. In this way, their social media use is both similar to and radically different from that of the feminists featured in the next chapter.

4

A Networked Feminist Killjoy Assemblage

*Social Media and Struggles against Sexism,
Misogyny, and Patriarchy*

In June 2013, while doing fieldwork in Beijing, I attended a performance of *Our Vaginas, Ourselves* (*Yindao zhidao* 阴道之道) (*OVO*, hereafter), a localized Chinese version of Eve Ensler's *Vagina Monologues* (see figure 4.1).[1] The show featured candid narratives about sexual assault and menstruation, but also humorous skits, such as a teacher leading a class on masturbation. Supported by *Yiyuan gongshe* (一元公社), a Beijing-based NGO, and organized by Bcome, a feminist volunteer group, it was performed by university students and NGO staff and volunteers.[2] Those who were involved in the production of *OVO* were part of a loose network of mostly young feminist activists, who were passionately committed to combatting patriarchal structures in China through performances and street art, workshops, trainings, and advocacy campaigns, all accompanied by the savvy use of social media. Taking part in *OVO* seemed to be a triumphant experience for the actors, imbuing in them a heroic feeling of breaking the silence on taboo topics. It was also moving and exhilarating for the audience, including me.[3] However, afterwards, when I mentioned the performance to several university students, some expressed deep interest, but others disapproved. Only five months later, when a group of young women who were students at Beijing Foreign Studies University (BFSU), as a promotion for their upcoming performance of *OVO*, shared pictures on social media of themselves holding signs with provocative messages such as "My vagina says I want freedom," they faced fierce pushback. As the photos spread across social media platforms, they generated a slew of comments, some supportive but most that called the students names such as "feminist prostitute" (*nüquanbiao* 女权婊, also translated as "feminist bitch"), derided their looks, and questioned their morality.[4]

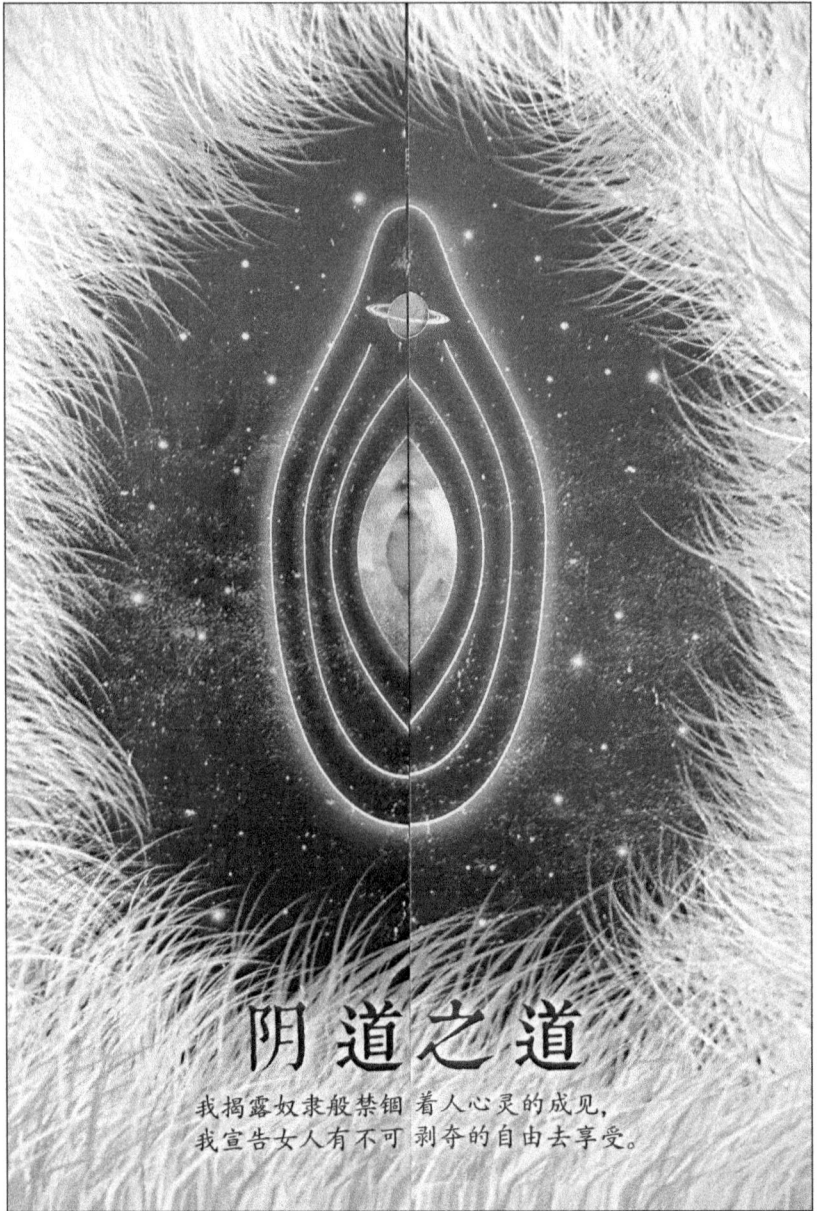

阴道之道

我揭露奴隶般禁锢着人心灵的成见，
我宣告女人有不可剥夺的自由去享受。

Figure 4.1. The cover of the playbill for *Our Vaginas, Ourselves*. Used with permission from Bcome.

The period in which I viewed *OVO* witnessed a blossoming of a new form of Chinese feminism, one that was independent and practiced by young women who were outside the official women's federation system and who were not afraid to use bold tactics to bring attention to negative phenomena affecting Chinese women. For example, just one year earlier, in 2012, several young women, called young feminist activists (*qingnian nüquan xingdongpai*, 青年女权行动派), or "YFAs," had engaged in a number of activities that called out sexism and misogyny, such as "Occupy Men's Toilets" (to bring attention to the paucity of women's toilets in public spaces), eye-catching street art/performances, and "photo contests," which were strategically designed to garner coverage by traditional mainstream media and to go viral on social media.[5] Their actions caused one Chinese journalist to officially declare 2012 the "Year of Feminism" in China (cited in Wei 2015), yet it should be noted that 2012 was the same year that "positive energy" became the top catchphrase on the Chinese internet (Sun 2012).[6] However, within two years of the *OVO* performance I attended, five of the young feminist activists were arrested and detained for 37 days, garnering international attention.[7] Two years after that, in May 2017, Song Xiuyan, the Party Secretary of the All-China Women's Federation (ACWF), the official mass organization that ostensibly exists to advocate for Chinese women, included "feminism" and "feminist supremacy" as part of a plot by "western hostile forces" to divide China.[8] A year later, the WeChat and Weibo accounts of Feminist Voices (*nüquan zhisheng* 女权之声), one of the most high profile feminist advocacy groups with arguably the most influential online feminist presence in China, were permanently shuttered after being temporarily suspended a year earlier. Since then, further harassment and persecution of feminists and those who identify as LGBTQ (there is much overlap) has continued, as has the denigration and censorship of feminism.[9]

In some ways, the official response to feminism in China is not completely surprising, as efforts to combat structural gender inequality by scholars, public intellectuals, NGOs, and activists have never coexisted easily with the official ideology of "equality between men and women" (*nannüpingdeng* 男女平等) that is grounded in a Marxist class analysis.[10] As well, while Song's accusation that feminism is part of a western hostile plot is a familiar trope in state rhetoric, the interaction of global and local feminists and feminist ideas have a long history in China.[11] The derision of feminists—as divisive, ugly, immoral, etc.—by wide swaths of the general public also aligns with patterns seen globally in both authoritarian contexts and liberal democracies.[12] In the United States, feminists are subject to online trolling and vitriol that Sarah Banet-Weiser (2018) argues is a result of "popular misogyny," or the status quo, largely invisible hatred of women that becomes visible, reactive rage when women are thought to be encroaching into areas where they don't belong (e.g., online gaming spaces, national politics, etc.).[13]

The Chinese context is different, however, as both historically and in the present, misogyny has been decidedly *visible* and undergirded by patriarchal Confucian ideas regarding the secondary role of women. Moreover, such official and popular enmity toward feminism has grown in tandem with the social formation I call neo/non-liberal China, one aspect of which is the overt propagation of conservative gender ideologies in state rhetoric and policy. The emphasis by Xi Jinping on women's role as wives and mothers is partly ideological and has historical precedent, but another reason is that China is facing a demographic crisis, to which the chosen solution is for women to get married and have babies.[14] These messages also circulate in commercial culture (see figure 4.2). However, long before the one-child policy and subsequent policies limiting the number of children that one could have were abandoned, and the role of women in the home became increasingly (once again) emphasized, entrenched sexism and gender inequality could be found in virtually all spheres of life: work, education, entertainment, family relationships, and so on (Fincher 2018, 2023; Liu and Moon 2018; Luo 2013; Meng and Huang 2017).

Simultaneously, a far more repressive political environment means that nearly any opposition to the gender status quo or attempts to call attention to "negative" phenomena, such as in *OVO* and through activist and advocacy campaigns, are severely constrained or completely suppressed (Xiong and Ristivojević 2021). Indeed, after deriding feminism, Song later declared in her speech that a major political task of the ACWF was to "guide women to obey the Party. . . . and to continue to nurture women with positive energy." Nonetheless, as the hostility toward feminism has increased, so too has the number of women who self-identify as feminist and use social media to spread knowledge, hope, and humor aimed at transforming China's regressive gender politics.

Figure 4.2. Packaging for dried black wood ear
mushroom. It is significant that the older two
children on the label are girls. Photograph by
Congshan He.

This chapter explores how a loosely networked group of young feminists dispersed across China engages with social media as a means of
envisioning and bringing into being a new ethics of gender and sexuality. I show how diverse experiences, feelings, attachments, and ethical
commitments brought the young women to feminism and how these
underlie their feminist praxis online, whether in the form of blog posts,
social media debates, or cultural productions designed to educate, entertain, and transform.[15] I focus in particular on their ordinary ethical

practices, or their exercise of practical reason and judgment as they work toward social change (Lambek 2010; Kuan 2015). I also highlight the role of affect and emotion in motivating and sustaining their efforts, or how they channel what Sianne Ngai (2005) calls "ugly feelings"—or "minor" negative emotions, such as anxiety, boredom, and irritation—as well as moral outrage, shame, deep pain, and alienation, but also optimism and hope, into transformative possibilities for themselves and others.[16] Like the BFSU students, for their efforts they are overwhelmingly stigmatized in the public sphere as the aforementioned feminist bitch/prostitute or "feminist cancer" (*nüquanai* 女权癌), which, I argue, are kindred spirits of the "feminist killjoy," Sara Ahmed's (2010) term for the feminist who spoils the happiness of others by pointing out gender inequality, sexism, and misogyny. In discussing their quest to challenge and eliminate various forms of discrimination and misogyny, however, I also, like Ahmed, want to emphasize the joy and pleasure this could bring them.

The young women's praxis, out of necessity, occurs primarily online, where they operate individually and in groups with shared interests and knowledge and in which members use digital technology to have a voice and remake culture, in the process forming something akin to what Mizuko Ito (2008) and danah boyd (2010) have termed "networked publics."[17] Rather than "publics," however, I call this formation a "networked feminist killjoy assemblage," which is an affective assemblage comprised of multiple articulations of bodies, discourses, emotions (anger but also joy), objects, technologies, and practices. As neo/non-liberal China is constitutive of the repression of any type of oppositional counterpublic, the term networked feminist killjoy assemblage emphasizes less its "publicness" and more its affective potentiality for action (Papacharissi 2015), heteroglossia (Bakhtin 1981), or many voices, and the value and effectivity of voice (Couldry 2010). In emphasizing the potential for voice to be valued, I am in no way implying that all feminists in China agree; on the contrary, there are quite contentious debates online about what a feminist is and what "real" feminism means.

In the years since I attended *OVO*, independent Chinese feminism has received much scholarly and popular attention, with particular focus on the strategies and campaigns of the young feminist activists mentioned earlier (Fincher 2018; Han 2018; Li and Li 2017; Wang 2018a, 2018b; Wang and Driscoll 2019), along with more diffused social media campaigns/hashtag activism, such as China's #MeToo movement (Han 2021; Xiong and Ristivojević 2021; Yin and Sun 2021; Zeng 2020) and COVID-19 protests (Yang 2022; J. Zhang 2023).[18] This body of work, some of it grounded in social movement theories of contentious repertoires or "connective" (rather than "collective") action, has contributed greatly to our understanding of individual and collective motivations, goals, and tactics. However, with few exceptions (cf. Zeng 2014; J. Zhang 2023), the role of affect and emotion has not been foregrounded, even though, as discussed in the book's introduction, in Chinese thinking, the heart and mind are not viewed as separate. Moreover, as I show, *gan* (感) "stirring and being stirred," is key in motivating the young feminists' actions to try to "stir" others. I thus seek to add to scholarship that has shown how affect, or "being affected, being moved," and emotion, linked etymologically to motion and movement, are central in social movements (Gould 2009, 2), perhaps even more so as they have become networked (Papacharissi 2015). I also highlight the specificities of these processes in a non-western, authoritarian context: neo/non-liberal China, where the state actively engages in strategies to move bodies in accord with its own desires by deploying a mode of therapeutic governance. Some strategies aim to generate "positive energy," for example, through channeling people's altruistic impulses into charity work (Ning and Palmer 2020). Others are coercive, directed at stamping out what is considered "negative," such as by criminalizing activism (Fu and Distelhorst 2018) and essentially "outsourcing repression" (Ong 2022) to non-state actors, including virulent nationalists who engage in their "patriotic" duty by harassing and trolling feminists.[19]

In this chapter, I include the pseudonymous voices of some well-known activists but also focus on unknown young women not operating in the limelight who nonetheless, as part of a broad network initially shaped by the more high-profile activists, are part of the "charged rhythms of the ordinary" (Stewart 2011, 446). In what follows, after providing background on the history of feminism in China since the early twentieth century, I discuss the affective and ethical forces involved in my participants' coming to feminism and how they situate themselves within the complicated and contested naming of feminism in China. I then discuss their use of social media and analyze three types of digital cultural productions, which have been underexplored in prior research: livestreaming, fansubbing, and video creation aimed at teaching, creating solidarity, and bringing laughter and hope to those seeking feminist connection and gender justice. Like the young creatives and domestic workers featured in earlier chapters, these young feminists' ethical commitments motivate their actions. Unlike them, however, many face quite negative consequences for claiming a feminist identity, openly expressing anger and discontent, and refusing to conform to the gender status quo. In the final section, therefore, I show how the women use social media to regroup and renew when they experience stress, boredom, backlash, and burnout.

A Brief History of Chinese Feminism

In the fall of 2015, while I was doing fieldwork in Beijing, several friends and associates were participating in numerous activities to celebrate Beijing +20, or the two decades since the 1995 United Nations Fourth World Conference on Women (FWCW) had been held in the Chinese capitol. All had been engaged in women's rights advocacy for a number of years, most of them through employment in official channels (universities, government think tanks, global development agencies, and government-organized non-governmental organizations [GON-GOs], such as the ACWF). The conference, and the accompanying NGO Forum that took place in a nearby suburb, had ushered in a sea change in participants' understandings of sexism, patriarchy, and gender and brought linkages with transnational feminism (Howell 2003; Wang 1998; Wang and Zheng 2010).[20] Afterwards, the Chinese government adopted the Beijing Declaration and the Platform for Action (a set of broad goals to promote women's empowerment and gender equality), giving those who worked on women's issues greater legitimacy and an accountability mechanism to push for state policies and resources to realize these goals.[21] Twenty years later, symposia and workshops mostly highlighted the progress Chinese women had made, through presentations and reports filled with statistics. These activities were organized and primarily attended by those "in the system," and as such, no matter how "radical" their own politics were, they had to toe the official line, at least in public. Outside of such official realms, however, many of these "inner" women's rights advocates, only some of whom identified as "feminist," acknowledged the serious issues continuing to plague Chinese women's lives.

The Beijing +20 activities offer an entryway to discuss the fraught history of feminism and ideas regarding women's empowerment and gender equality in China. Several recent scholarly articles, monographs, and edited volumes on Chinese feminism have given detailed accounts of this history.[22] Below, I focus on three aspects that are important for understanding the ethical commitments and feminist practices of the young women featured in this chapter: first, the manner in which gender ideologies have shifted over time in accordance with the party-state's prerogatives; second, global linkages, academic exchanges, and cultural flows in the 1980s and 1990s, such as the FWCW, which opened the door to my participants' future exposure to feminism (in university classes, performances such as *OVO*, etc.); and third, informal/unofficial entities that emerged in the late 1990s and early 2000s to critique patriarchy, sexism, and misogyny and eventually used digital media to offer a feminist vision and engage in feminist action.

Feminism entered China in the late nineteenth/early twentieth centuries when the country was in the throes of political, social, and cultural turmoil after it had suffered numerous humiliations at the hands of foreign powers (particularly the Japanese and British). Although some Confucian reformers had advocated for women's emancipation, the New Culture and May Fourth male intellectuals/reformers viewed the "backwardness" of Chinese women, who were largely illiterate and taught to follow the Confucian "three obediences and four virtues," as a metaphor for the weak nation. Thus, they looked abroad for ideas they thought could help strengthen and modernize China.[23] With the introduction of translations of works on western feminism, new concepts and vocabulary appeared, including "equality between men and women," "women's liberation" (*funü jiefang* 妇女解放), and "women's rights/power" (*nüquan* 女权) (Sudo 2007; Wang 2017).[24] The Chinese translation of "feminism" (*nüquanzhuyi* 女权主义) circulated widely "for its explicit association of '-ism' with 'women's rights or power'" and as a means of combatting "feudalism" (Wang 2017, 3–4).[25] A new "modern" word for "female sex" (*nüxing* 女性) also emerged in the 1920s to signify universal woman (Barlow 1994). However, although there were a small number of women reformers, in most cases the underlying goal of male-dominated nation building was that liberated and educated women would produce better future citizens.

In 1949, the People's Republic of China was established by the Chinese Communist Party (CCP) under the leadership of Chairman Mao Zedong. Many male CCP cadres had always eschewed *nüquanzhuyi* as "bourgeois" due to its association with imperialism and capitalism and because they believed it divided men and women (Wang 2017).[26] The Marxist belief that women's emancipation would be achieved through their participation in production (as opposed to reproduction) was adopted, and gender as an analytical category was subsumed by class, which became the key lens for understanding social relations. From its inception, the state was committed to the aforementioned "equality between men and women."[27] To realize this commitment, the ACWF was established early on; however, it had a marginalized position in the larger Party structure, and its primary purpose was to lead and mobilize women to carry out Party policies under the banner of "women's work" (*funü gongzuo* 妇女工作), even when such policies did not necessarily benefit women (Jacka 1997; Wang 2017). Maoist radicalism during the Cultural Revolution (1966–1976) resulted in the seeming further liberation of women, with slogans such as "women hold up half the sky" and "everything a man can do a woman can do, too" embodied by the strong and daring Iron Girls.[28] In reality, there was a "gender erasure" as women had to conform to men's hairstyles, clothing, speaking style, and other mannerisms (Yang 1999). A paradox of the state's, and by extension the ACWF's, approach to gender equality at this time was that, while women were integrated into social, economic, and political life, their position was largely dependent upon state-imposed structures, or "state feminism."

After China's reform and opening policy began in the late seventies, there was massive restructuring of all aspects of life, and the safeguards for women embodied in state feminism began to erode. As Wang Qi (1999) notes, once these structures were gone, and "the state's role changed from transformative to operative, women not only had to look elsewhere for ideological support . . . , they also had to confront the old social ethics that reemerged in the retreat of state ethics" (29). The resurfacing of patriarchal notions of gender (which had never actually gone away) manifested in the elevation in state and popular discourses of notions of binary, "natural" gender as a means of repudiating the former gender erasure (Evans 1997; Yang 1999). Such ideas were quickly bolstered by China's fledgling commercial media, which commodified and sexualized women's bodies, and, along with state media, regularly featured discussions about ideal (e.g., traditional) womanhood (Evans 1997). In work, education, and politics, women faced forms of discrimination that were assumed to have been eliminated.

Starting in the mid-to-late 1980s, local women's federation cadres established associations and women-focused NGOs dedicated to researching and intervening in what they perceived to be the alarming negative effects of the reforms on women (Milwertz 2002; Wang 1998).[29] At a few universities, small departments and salons for research on women were also established. These efforts were influenced by international exchanges in which Chinese scholars and those doing "women's work" were introduced to western, primarily liberal, feminist ideas and Chinese translations of feminist (mostly French- and English-language) texts, even if these ideas did not always align with China's socialist and pre-socialist history (Shih 2005; Wu 2005). While some critics argued that the embrace of binary gender and traditional gender roles was the result of a masculinist discourse and commercial culture that helped revive blatant sexism, Li Xiaojiang, the founder of women's studies in China, disagreed. She contended that these changes were positive to the extent they could "awaken" Chinese women's suppressed "female consciousness" (*nüxing yishi* 女性意识), which had been denied them in previous decades (Li 1999).[30] In these burgeoning exchanges, feminism was often translated as *nüxingzhuyi* (女性主义), or woman-ism, rather than *nüquanzhuyi*, or women-powerism, a detail to which I will later return.[31]

A key turning point came with the aforementioned FWCW in 1995. In the wake of the conference, more partnerships and NGOs developed, often with funding from international development agencies, to carry out projects aimed at improving the lives of women (Howell 2003; Wang 1998). Wang Qi (2018b, 262) calls this time a fruitful period of "post-Mao bottom-up feminism" in women's studies departments and "NGO feminism" in China's fledgling civil society.[32] Even so, many women who attended the conference and then continued or began working in GON-GOs and various ACWF projects (poverty alleviation, rural women's development, women's political participation in villages, etc.) afterwards did not and do not necessarily identify as feminist for various reasons, some strategic and some ideological.[33]

In the relatively open period that followed the FWCW, it was possible for those who were dedicated to fighting systemic gender inequality, whether they worked for the ACWF or were loosely affiliated, to put their ideas into new forms of action and advocacy. These included the Women's Media Watch Network, also called the Media Monitor for Women Network (*Funü chuanmei jiance wangluo* 妇女传媒检测网络), and the Anti-Domestic Violence Network (*Fanduijiating baoli wangluo* 反对家庭暴力网络), established in 1996 and 1998, respectively. The former was created by journalists who worked for *China Women's News* (*Zhongguo funü bao* 中国妇女报, an ACWF publication) to amplify women's voices and push for gender equality in the media (Cai, Feng, and Guo 2002). In 2009, the e-newsletter Women's Voices (*Nüsheng dianzibao* 女生电子报, also called Gender Watch Women's Voice) was founded by Lü Pin.[34] This publication was unique in taking a firmly feminist stance when examining issues that primarily affected women, including sexual violence, work and education discrimination, childbirth and sex-selective abortion, body image, and sex work, among others (Women's Media Watch Network 2009–2011).

Eventually the name Women's Voices was changed to Feminist Voices, which began its microblog presence on Weibo in 2011. This year is significant, as it was the same year of the Wenzhou train accident (mentioned in the book's introduction), when Weibo was still relatively new, and many urban, educated Chinese had high hopes that it would become a sustained means of forcing more transparency and accountability on the part of the government and provide a "public square" for the exchange of ideas. Although the period from roughly 2009 (when Weibo launched) to 2013 is known as the "Golden Age of Weibo," the commercial nature of Weibo meant that many of the most influential users—movie stars, singers, and athletes—gained (and still gain) their appeal through relying on tropes of essentialized binary gender. Indeed, during this time, two of the most popular people on Weibo (in number of followers, or "fans") were Sora Aoi, a former Japanese porn star, and Liu Jishou, who gained over 500,000 fans through rating the appearance of men and women who sent their photos to him. His comments were usually profanity-laced, always snarky, and often deployed misogynistic language and/or drew upon gender stereotypes and urban prejudices against rural residents.[35]

In this mixture of an increasingly sexualized and commercialized culture, combined with entrenched institutionalized patriarchy, Feminist Voices perhaps had the most profound influence on the new feminist activism that emerged in the 2010s—indeed all of my informants cited its significance. However, the earlier interventions by women's studies scholars and media practitioners, including feminist film director and activist Ai Xiaoming's first adaptation of *The Vagina Monologues* on university campuses in 2003 (the inspiration for the version I viewed ten years later) along with many other efforts, laid the foundation for such activism, as some have pointed out (Fincher 2018; Wang and Driscoll 2019; Xiong 2019). It was during this flourishing of ideas and exchanges dedicated to issues of gender, women's empowerment, feminism, and in some cases sexuality, that the high-profile YFAs and other young feminists came of age.

Becoming Networked Feminist Killjoys

In the class, there was one chapter on feminist critique, so I got to know about feminism from it . . . and I just thought, "Wow!" It just opened up a new world for me. . . . It helped me understand why the world is like this . . . I got really interested, and I wanted to systematically learn something about feminism, so I read every book that's related to feminism or gender studies in the library when I was at my university, but you know there were so few (laughs). Later I went to XX to study gender studies, but I was kind of disappointed there because I was trying to find some friends who are also quite interested in gender issues and maybe want to do something, but I really didn't find any classmates that are so passionate like me [about] changing things. When I was in XX during 2010 and 2011 Weibo was quite popular . . . and really active. So, I tried to find some information on feminism in China, and then I found Feminist Voices.
—Hai, born in 1988[36]

The quote above was spoken by Hai as she recalled her journey to becoming a feminist. Although the young women in this chapter had different backgrounds and were born across a thirteen-year timespan, all of them, like Hai, expressed deep emotions while recalling particular experiences that left them searching for something to provide meaning for "why the world is like this." Finding this meaning in feminism was like turning on a lightbulb—Hai's "Wow!"—that then spurred them to want to "do something" and connect with like-minded others. In Hai's and others' narratives, the central role of *gan*—to stir and be stirred—as it was articulated to an ethical construction of the self, became clear. To unpack this entanglement of stirrings, emotions, objects, and ordinary ethics in their identification with feminism, below I discuss the experiences the young women shared that were key in their joining a networked feminist assemblage, in which voicing anger and discontent—and turning away from the objects that were supposed to make them happy in order to find happiness elsewhere—rendered them killjoys or feminist cancer in the eyes of the party-state and others who viewed feminism negatively.

Like many young feminist activists in China, several of these women were born in urban areas. Many were singleton daughters and had relatively privileged upbringings because they were not competing with siblings, especially brothers, for resources. Later experiences of discrimination in the "real world" thus came as a shock.[37] Emily, for example, articulated very clearly her happy childhood and later disappointment saying, "I was . . . maybe spoiled by my parents, and I didn't suffer a lot of gender discrimination. [They] always championed me, so when I faced a setback or failure or was treated unequally outside of my family, I would be really angry."[38] Ning, who had a similar background, had learned as a child that "women hold up half of the sky" and acknowledged the resources poured into her. However, although she worked hard and did very well in school, as an adult she faced job discrimination. "The ideology is backwards," she said. "It gave us a huge frustration."[39] Xiao Wei as well was raised learning that boys and girls were equal, but after graduating from college she was passed over for a job, presumably because her employer assumed she would get married and have a baby. A journalist added that "the real struggles or the moment of revelation comes *after* you've graduated. Because . . . especially if you're from a first tier or second tier city in China, you're brought up believing you are equal to men. Because we come from a single child generation."[40]

Others, however, were not as fortunate. Ting said she witnessed "violent events between my parents before, and I thought it wasn't okay, but

I never understood why this happened. Later I learned this is because of patriarchy."[41] Fang Ai, who was from a small rural town and had a younger brother, described growing up in a very "*zhongnanqingnü*" (重男轻女) family, meaning a patriarchal household where boys are favored and girls are undervalued.[42] Her father called her "dumb," and he constantly reminded her that he was going to bequeath everything to her brother, which caused her stress. Speaking of these experiences she said, "It's not fair. I've been really hurt."[43] Another participant, Xiao Yan, felt depressed because she was attracted to women but did not have the language or environment to express her sexuality. Still another had been sexually assaulted by her uncle at a young age but wasn't believed when she told her parents.

The young women's accounts accord with research that has tracked barriers to Chinese women's participation in the labor market (Brussevich, Dabla-Norris, and Li 2021) and, in the case of Fang Ai, with the fact of lingering son preference in rural areas. They also reveal how a range of experiences and emotions, and feelings of deep disappointment, anger, hurt, shame, and/or depression caused by injustice or violence due to their gender and/or sexuality, left deep impressions on their bodies/minds. When these young women then got to college and took a gender studies class, or even one unit of a course that exposed them to feminism (many cited reading Simone de Beauvoir's *The Second Sex* as akin to removing the blinders from their eyes and helping them understand the social construction of gender), or viewed a performance of *Our Vaginas, Ourselves* on their campuses, these "objects of feeling" stirred something inside of them (Ahmed 2014).

Feminist scholars early on documented and theorized how embodied passions, particularly anger that is both individual and political (i.e., moral outrage at unjust structures), are instrumental in feminist praxis aimed at social transformation (Lorde 2007; Spelman 1989). Historically in China, however, negative emotions such as anger and depression have been individualized and often attributed to incorrect thinking rather than social structures (Yang 2016). In the neo/non-liberal present, discourses of "positive energy" continuously circulate through official rhetoric as the state harnesses affect as a mode of control and to motivate productivity, e.g., happy citizens make "good" citizens, "good" here meaning compliant, entrepreneurial, and patriotic (Lindtner 2020).

Ordinary people also invest in an array of self-help regimes to try to find happiness as part of the "psycho-boom" within popular culture (Kleinman 2010; Y. Zhang 2018; Zhang 2020).

Young women turning to feminism to find happiness and meaning should seem to align with these popular trends. However, another crucial aspect of neo/non-liberal China is shored-up patriarchy, not only institutionalized in labor, education, and family structures but also boldly proclaimed by Xi Jinping and other state entities as part of the Chinese Dream. As just one example, the "morality schools" that teach "feminine traits" mentioned in the book's introduction find a counterpart in ACWF-sponsored courses in different regions that similarly train women to be "proper ladies" (Fincher 2018, 169). For some young women, the reality of sexism, which completely contradicts the messages that were inculcated in them during childhood regarding women's equality, compel them to speak out. Their refusal also to turn a blind eye to other negative phenomena—misogyny, rape culture, etc.—disrupts a state-sanctioned version of harmony and positivity. In the state's formulation, then, and among a large segment of the population, *feminism* is the cause of one's discontent, disharmony, and depression. Indeed, feminism itself is viewed as akin to a malignant disease: cancer. Like Ahmed's (2010) "feminist killjoy," who does not find happiness in what is supposed to make them happy, these young women, not only those who identify as queer, in addition to refusing to be silent also often do not go along with conventional norms of gender and sexuality. For example, they reject marriage and motherhood, which historically in China was unheard of. In a similar manner, Ahmed's (2010) "unhappy queer" can be both dissatisfied (unhappy) by the normative choices offered them, or cause others, such as parents, to be concerned that their decision to lead a life not centered on heterosexual partnering will lead to their unhappiness. Even today, it is considered a betrayal of filial piety when children do not honor and obey their parents by getting married and having a family.[44]

Such a turning away from normative constructions of happiness and gender role expectations is thus not only attributed to young feminists' anger, but also to their supposed selfishness. Selfishness has been seen as a key attribute of members of the "post-80s" and "post-90s" generations (those born in the 1980s and 1990s), who grew up when previous ideas about "serving the people" and the primacy of the socialist collective were replaced by an emphasis on materialism and individualism (Rosen 2009; Yan 2010).[45] Yet the young women featured in this chapter had a different take on this assumption, as perhaps best summed up by Emily during our interview, in which we discussed the differences between her generation and the state feminists (*guojia nüquanzhuyizhe* 国家女权主义者, *zhe* here equals the suffix "ist" in English) who came before her. Such generational differences have in some cases created rifts between younger and older feminists. In contrast, Emily, who was involved with various NGOs, expressed admiration for their prior efforts on behalf of women, but also a different starting point:

E: [Their work] was for a bigger cause. It was for the nation's benefit because the nation wanted women to be laborers, to strengthen and develop our country, so they offered women some benefits. But I think the [state feminists] really wanted to help women. Although they relied on a nationalist hat [*meaning they relied on the framework of nationalism*] they did a lot to help women. And they used another word, women's liberation. I think this word can show feminism's real meaning, but it's misused, so. . . .

C: How's that different from you and your peers?

E: Because our generation is more focused on ourselves, like individualism (*gerenzhuyi* 个人主义).

C: Is that okay?

E: It's okay. It gives us more—a space to consider why we need a country or a collective or group [and] why we can't just represent ourselves and fight for ourselves.

C: So, you say it's about yourselves, which sounds very selfish, but what you're doing doesn't sound selfish.

E: But I think selfishness is a good thing for women. Because women were taught to give so much. Yeah, I think some selfishness is good for us . . . [Feminists] fight for more women, but the starting point is ourselves. That is to say, we are like this. And I say this to others, "If I do a (feminist) action, an important part is for myself."[46]

While acknowledging her generation's so-called "selfishness," Emily insisted this focus on the self was ultimately an ethical project of the self, with implications for improving the status of all Chinese women. In this way, she set herself and her feminist (*nüquan*) peers apart from "fake" or "pseudo" feminists (*weinüquan*, 伪女权) and "pastoral feminists" (*tianyuannüquan* 田园女权), who, in popular discourse, only see feminism as a means of acquiring personal benefits (Peng 2020; G. Zhang 2018).[47] Other women also described an underlying higher purpose, drawing on the very socialist morals that have supposedly been discredited in the present era. For example, Hai stated, "When I was quite young, I thought that people should not live because of . . . just to make a living, but also maybe to pursue some higher ideas. . . . When I was young, we still had some of that socialist education," meaning they were taught altruism, sacrifice, and living for a greater cause.[48]

Embracing with a passion feminism, which is stigmatized and marginalized, but finding few likeminded people around them, as in Hai's quote earlier, meant that all of the participants in this chapter searched for connection online. Although early in the 2010s, Feminist Voices was one of the few venues for such connection, over time more spaces developed, resulting in a dispersed network. Like Feminist Voices, some of these more public arenas were later shut down due to increasing censorship and repression, but other less public nodes persist. Whereas in the past, feminism thrived in small interconnected off- and online spaces, in the neo/non-liberal present the latter arena is much safer, even though it comes with its own risks. In the remainder of this chapter, I discuss various manifestations of this networked feminist killjoy assemblage. Again, this term is meant to capture the articulations of affect, ethics, digital media, feminist praxis, and young women in a loose, relatively small (compared to the general population) configuration of decentralized nodes. I highlight the women's dialogue regarding the meaning of feminism, creation of feminist knowledge and community, and renewal as they countered state-sanctioned and popular misogyny, which in China often become entangled when government hostility to feminism, "a western hostile" force, stirs up rage in others who see feminism as anti-China and as perpetuating hate. As feminist killjoys, they did suffer pain and anxiety at times, but the "joy" in killjoy, a feeling based in a belief that their actions had deep meaning, and the pleasure that came by being involved with feminism, were also part of their experience.

What's in a Name? "Feminism" as Contested Terrain

In a 2021 English-language volume on feminism in China, which includes contributions from established Chinese women's rights activists and scholars, the editors use a word adopted by women's studies and literary scholars in the 1980s and 1990s, *nüxingzhuyi*, translated variously as "women-ism" or "feminine-ism," and call it a "broader category that can incorporate" *nüquanzhuyi* (Zhu and Xiao 2021, 1). However, in the same volume, *nüquanzhuyi*, which is akin to "women's rights/power-ism," is given as the term used for feminism by young feminist activists today (Wang 2021, 139). Indeed, in prior scholarship on feminist activists, it is often taken as a given that they use *nüquanzhuyizhe* to describe themselves. Nonetheless, as I discuss below, the young feminists making up this particular networked assemblage had diverging ideas on which term was best, and their views were deeply connected not only to their identity but also their beliefs about how best to stir/move others. Moreover, contrary to popular belief, except in one case there was broad acceptance of differing perspectives based on the understanding that ultimately they were all working toward a common goal. I should note that these are but two terms for feminism in China, and new words for different types of feminism seem continuously to appear.[49]

"Nüquan *has already been demonized* (yaomohua 妖魔化)"

Although they were in the minority, some of the young women preferred the term *nüxingzhuyi* for two reasons. First, it aligned with their attachment to a sort of transnational lifestyle/popular feminism. Second, because they were committed to "spreading feminism," they saw it as less threatening and thus as a means of engaging others. Jennifer, who was 27, queer, and single, identified as *nüxingzhuyizhe*, because, in her words, it was "less extreme," and the concept was "evolving."[50] Lana, who had studied and worked in Beijing and Canada, said she used *nüxingzhuyizhe* because the term is more "moderate and not that direct. . . . Nüxing* is more elegant and will be easier to be used to educate more people . . . *nüquan* has already been demonized (*yaomohua*)."[51] Lana attributed her time living abroad to opening her eyes to unequal gender politics. While at her first job in Beijing, she had never questioned why, as the only woman on her team, she had to serve tea at meetings. In Canada, however, she began observing how women seemed to be more independent:

> "Gradually I thought that if people kept a very Chinese traditional . . . gender mindset, they would be very upset when they experienced something complicated or pressure, but if you think in a more feminist way it will benefit you. It will make you happier . . . and it will just make you stronger."

The "pressure" to which she referred above affects both men and women. "Men have pressure to get good jobs" so they will be eligible bachelors, she said, and young women have tremendous pressure "to marry and have children." Both single and in their late twenties, Lana and Jennifer were subject to the "leftover woman" (*shengnü* 剩女) discourse, a term for single women over 27, especially educated, career women, viewed as beyond the desirable age for marriage.[52]

Both Lana and Jennifer used social media to engage others in feminism. Jennifer had a public WeChat account in which she discussed events abroad related to gender and sexuality and explained them to a Chinese, middle-class, urban audience. Her posts included discussions of Hillary Clinton's presidential run and Australia's legalization of same-sex marriage, among other topics. Lana frequently posted about her discontent with gender representations in Chinese popular culture. For example, after rewatching the Chinese movie *Ex-Files* (*Qianren gonglüe*) and writing a critique of the gender dynamics depicted in the film, she ended her post with "Just wondering: am I living [now]?" A friend quickly commented that she was "too progressive." In response, Lana said she debated with him, but "then I gave up," noting the futility of such a debate. "Giving up" did not mean no longer posting, it just meant she refused to expend energy in a "useless battle."

Both Lana and Jennifer could be accused of being feminists only for their own self-interest and not because they wanted to fight for structural change. In many ways, their feminist engagements on social media reflected a middle-class "popular feminist" perspective: it was consumerist, individualistic, and visible (Banet-Weiser 2018), yet they did question social structures.[53] In some ways, it was no more individualistic than Emily's statement that the starting point of feminist activism was herself. Lana, Jennifer, and the few other participants who preferred *nüxingzhuyi* saw it as a strategy to bring more people on board as feminists. However, even though they were not activists (and not all of those who identified as *nüquanzhuyizhe* were either, as I discuss below), they still faced pushback. In mainstream discourse, they were a "bitch/prostitute" or a "cancer." In being "willful" subjects (Ahmed 2010) and turning away from what was supposed to make them happy (marriage and family) and finding freedom, strength, and happiness through feminism, they embodied the feminist killjoy.

In China, as elsewhere, feminism is an object that accumulates affective value, and as Ahmed (2014) notes in her discussion of why she prefers to focus on objects over feelings, any guarantee that a shared feeling will circulate between different bodies is nil. During interviews, many women noted that putting the words "rights" and "women" together scares off some people, especially men. However, there was a general feeling that it did not matter how nice one behaved or how much one smiled, "people will hate you anyway" if you are a feminist. Thus, the common sentiment was "you might as well call yourself *nüquanzhuyi*." Indeed, memes depicting feminists as ugly, man-hating, and abnormal flow across Chinese social media. For example, figure 4.3a is a meme (from Weibo) in which a young woman is saying, "I'm 1.68 meters tall, weigh 96 *jin*, have 85-60-88 [cm] measurements, my monthly salary is 50,000 yuan/month, my husband loves me. Why would I participate in feminist activism [represented with the character for female and a fist] and curse on Weibo every day?" She is speaking to a woman that is supposed to represent the ugly, unsuccessful, angry feminist. Feminists also counter with their own memes, as in figure 4.3b (from Douban), in which enoki mushrooms represent small male penises, and the text reads, "Shut up you guys."[54] Both memes rely upon social constructions of masculinity and femininity to make their point, but evoking anxieties about men's virility is meant to especially sting.

Figures 4.3a and 4.3b. An anti-feminist meme and a meme telling men to "shut up." Screenshots by Qingyu Dai.

你们闭嘴

Figures 4.3a and 4.3b. (*continued*)

Nüquanzhuyizhe *as Refusal and Shared History*

Many of the young feminists featured in this chapter proudly pro-claimed themselves as *nüquanzhuyizhe*, a term that gained much wider usage after 2012 due to its association with the high-profile feminist activists mentioned earlier. The young women's reasons for adopting the term were varied but also overlapping, gesturing to refusal, history, and solidarity. For some, identifying as *nüquanzhuyizhe* was a distinctly oppositional stance. Ting, who had previously used *nüxingzhuyizhe* while in college, later found the term too weak. In her words:

It's like *Lean In*, you know? Kind of like you're in a privileged circle. You can use *nüxingzhuyi* so you won't irritate other people. Then your rights or your privilege won't get disturbed, although as I'm saying this [it sounds] really mean (*laughs*). . . . Before you came out [as a feminist] you were *nüxingzhuyizhe*. After you came out, you are *nüquanzhuyizhe*. Your identity is "I'm offensive. I bite sometimes."[55]

By declaring she was "offensive," Ting proudly embraced her feminist killjoy identity, channeling anger into action. This stance also recognized the gendered structures that often get ignored or are tacitly accepted in what she called "Lean-in" feminism, referencing Sheryl Sandburg's wildly popular, but also widely critiqued, book.[56]

Others adopted *nüquanzhuyizhe* to link to a shared history, before the CCP-era of state feminism. Ace, an activist, stated, "When you say 'quan,' it's more like we are heirs of the feminist workers of the Republic of China before the early Communist Party."[57] Use of this term, then, rejected the "men and women's equality" discourse of the party-state, implicitly critiquing it for failing women. As a political stance, such usage also linked to human rights discourses that emerged in tandem with feminism during the Republican era, as discussed earlier. In her work examining feminist and queer activism in China, Jia Tan (2023) translates *nüquan* as "rights feminism" to highlight its connection with the emergence of specifically rights-related practices (while also acknowledging that using this term is not necessarily always related to activism around rights). Julie, a university student, also adopted *nüquanzhuyizhe* for this reason, stating, "I consider myself a *nüquanzhuyizhe*. . . . [which] is more about rights, especially political rights, economic rights for woman. I focus more on rights we deserve."[58] Thus, identifying as *nüquanzhuyizhe* was a political stance. Xiao Yan said, "We say we are *nüquanzhuyizhe* because it's more direct and we want to change the situation that it isn't a good word. . . . On social media when we use this term *nüquanzhuyizhe*, people attack these feminists."[59] A hallmark of neo/non-liberal China is that, even more than during prior regimes of the last two decades, any public discussion or action in support of human rights, whether connected to feminism, labor, property, or something else, has become fraught with risk. Even loose association with such ideas can lead not only to personal attacks and online and offline harassment, but also to detention and potential long jail sentences.

Still, other young women who did not identify as activists also used *nüquanzhuyizhe* as an ethical stance; that is, to draw people into a feminist network that could help them and make them happy. Jia, who co-moderated a feminist WeChat group, had never been involved in street performances or other offline activist campaigns because she did not believe they could work in China, not only because of censorship, but also "because the public doesn't really . . . agree with that."[60] The purpose of her group was to bring people closer to feminism. "I like people to discuss, and I also like them to invite new people to come and enjoy the group, especially, like, high school students . . . so that we will expand to [having] more feminists," she said.[61] Her feminist aspiration stated here thus shared much with Lana and Jennifer's reasons for using *nüxingzhuyizhe*.

All my informants said they preferred one identification over the other, usually *nüquanzhuyizhe*; however, most were not concerned if someone used the opposite term. In Ace's words, "I prefer *nüquanzhuyizhe*, but if someone uses *nüxingzhuyizhe* I don't have a problem with that."[62] Sue, who identified as *nüquanzhuyizhe*, stated, "People like to say bad things about *nüxingzhuyizhe*. . . . [But] they are the same thing because they try to solve the same problem—to promote gender equality. There is not a difference but just different approaches to solve the issue."[63] Fang Ai, whose name on one of her social media accounts roughly translates as "Righteous Anger," disagreed, however, saying, "I don't think it's okay to call yourself a *nüxingzhuyizhe*. If you want to fight, fight." Echoing some of the sentiments above, she continued, "Because some feminists are very radical, people don't like them because of that, so they use *nüxing* to sound softer and speak more gently, but that doesn't help. It doesn't make men open to feminism or sympathize with you. They just don't care about you."[64] Still, despite Fang Ai's feelings that using *nüxingzhuyi* basically capitulated to misogyny, most women featured in this chapter believed it should not be written off entirely because it, too, does political work.

Feminist Cultural Production

The way these young women identified as feminists informed their feminist praxis. Although Chinese feminists of all stripes create blogs, podcasts, memes, hashtags, and the like (Peng 2020; Yang 2022; Zeng 2020; J. Zhang 2023), as mentioned earlier the YFAs' advocacy has received much attention in academic scholarship. Reasons for this include how, at least initially, they intentionally sought to be newsworthy to capture attention (Li and Li 2018); how their tactics represented "a new repertoire of contention" in China (Wei 2015); how they strategically used social media for hashtag activism (Wang and Driscoll 2019; Xiong 2019); how the campaigns they started in some cases led to tangible policy changes (Tan 2023; Wang 2018b); and how they faced misogyny and eventually repression and ongoing surveillance and harassment (Fincher 2018; Han 2018; Wang 2021).[65] Their bold and creative feminist practice has been transformative and has inspired many in and outside of China, yet they have also been subject to the critique that they promote a western, liberal notion of feminism that is not radical enough (Xue and Rose 2022).[66]

To add to our understanding of young feminist praxis in neo/non-liberal China, here I discuss three feminist cultural practices that have been less remarked upon: livestreaming, fansubbing of foreign media, and video production. This work has been (and still is) done by individuals, small groups of friends, and NGO staff and volunteers, who, for the most part, are not operating in the spotlight. It appears across diverse networked platforms and carries a range of themes. Some of the subject matter is aimed at advocacy, while other content is meant to educate and/or entertain. Like the young creatives discussed in chapter 1, a deep stirring within moves these young feminists to engage in cultural production. However, whereas for many of the young creatives, self-transformation rather than societal transformation was their primary motivation, for the young feminists it was the inverse. Although the starting point was the self, in Emily's words, the larger goal was to move others (generate Hai's "Wow!"), change their hearts/minds (*xin*), and draw them into this networked feminist killjoy assemblage. In my analysis, for the sake of space I focus on one representative example of each.

Livestreaming

Fang Ai, whom I mentioned earlier, was one of the most passionate young feminists I knew. At the time we met, she was in her early twenties and studying to enter a master's degree program while working part-time creating content for a public WeChat account focused on legal issues related to gender equality. Fang Ai felt isolated as a feminist, stating, "In real life . . . there's not a lot of people [who identify as feminist]. People think you are strange if you talk about feminism."[67] Like others with marginalized identities, she looked for community and validation online, initially turning to Weibo to find key opinion leaders (KOLs) such as Li Sipan, a well-known feminist activist, and Li Yinhe, a public intellectual and sexologist. Later she joined some WeChat groups devoted to sharing and discussing feminist issues.

Fang Ai was extremely dissatisfied with the representation of gender in China's popular culture. "As a feminist, when I watch Chinese movies or television shows, I feel offended to my core. This feeling is terrible," she said, her voice sounding distraught. Indeed, the sexism and misogyny in China's state and commercial media have been the target of feminist protest (Xiong 2019). China's livestreaming industry as well is known for an abundance of female livestreamers who either perpetuate very traditional notions of womanhood and femininity or engage in what has been labeled the aforementioned "fake feminism" that critics say is actually about their own selfish interests.[68] Because the emphasis of the platforms and the livestreamers is on monetization, these representations are not surprising. Women hosts who do not conform to such hyperfeminine standards are disciplined in particular ways (Zhang and Hjorth 2019).

In this context, like young feminists elsewhere, Fang Ai channeled her anger and disappointment into action (Kuo 2019; Mendes, Ringrose, and Keller 2019). She had participated in prior hashtag activism but was moved to do something more, so she created a livestreaming (virtual) room on Bilibili, a hugely popular video sharing and livestreaming platform. Her motivation was to alleviate her and others' "despair" and to bring hope instead. As such, her livestream focused not on generating income but rather a higher purpose—educating and building feminist community.[69] She carefully chose different topics, for example, gender as a construct (in her words, "Like, girls must wear pink and boys must wear blue. . . . I think it's ridiculous."). She also critiqued patriarchal structures, which she had experienced as a child and which had left a deep impression on her, as mentioned earlier.

As a result of Fang Ai's livestream, several viewers added her as a WeChat contact, which led her to create a WeChat group in which members discussed feminist issues. A prominent difference between a

platform like Bilibili and YouTube is that the former is even more focused on community, which is created in several ways—viewers can
subscribe to a livestreamer's channel, make comments, send virtual gifts,
and type *danmaku*, or "bullet comments," which flow across the screen
during a livestream.[70] The number of members (50) in Fang Ai's WeChat
group was small by Chinese standards, but maintaining the group and
the livestream required a large amount of affective labor on her part. As
she stated, so many young Chinese women "feel it's very difficult, and
I wanted to encourage them." Such encouragement was mutual in the
supportive space of the virtual group.

Raising issues that brought up anger, sadness, and hurt meant that in
the state and others' eyes, Fang Ai was perpetuating negativity; hence,
she was a feminist killjoy, even though her purpose was to generate
healing and hope. As the government increasingly cracked down on
livestreaming platforms, Fang Ai was censored by Bilibili. One way that
Bilibili has sought to comply with stringent state-mandated content restrictions, such as the requirement to promote "mainstream values," is
to create a "Disciplinary Committee" that taps into the platform's community ethos and users' sense of agency in order to involve them in
content regulation (Chen and Yang 2023). Although these measures can
prevent harm, they also potentially give license to create harm, such
as when some users find content to be "offensive" because it is not in
line with their and the party-state's views, in this case regarding ideas
about gender, feminism, sexism, and the like. Discussing her experience,
Fang Ai said, "If you talk about feminism with the audience during the
livestream, they ban you. Because it's really strict." This tightening space
made her feel like "there were threats around me that I couldn't see, and
I felt insecure, so I had self-awareness to stop talking about the issue
because I knew if I kept it up there would be problems. So, I gave up my
livestream."

Feminists globally have experienced threats, doxing, trolling, and harassment (Ging and Siapera 2019; Mendes, Ringrose, and Keller 2019).
In neo/non-liberal China, individual users are encouraged to engage
in such behavior by the party-state, and it is enabled by the platforms'
rules and affordances (Liao 2024). Furthermore, such online threats
and harassment can be accompanied by additional offline forms of in-
timidation by state entities, including visits to one's home by police and

being invited to a police station to "drink tea" (a euphemism for being interrogated). Some women I knew also lost jobs and/or housing due to pressure from above on their employers and landlords. Those who had experienced such phenomena or knew others who had called the online realm increasingly "scary," "violent," and anxiety-inducing for feminists. Still, Fang Ai continued to maintain her WeChat group, as an ethical commitment to uplift those in the group and to try to change the gender status quo, even as the emotional toll and the threat of censorship became greater.[71]

Fansubbing

Fansubbing is the practice of translating foreign content and adding subtitles (in China, it was initially done with movies and television shows but more recently also with online lectures, courses, and the like). Chinese fan communities emerged years ago when pirated DVDs of foreign (especially Japanese- and English-language) media content became readily available. As the internet dramatically increased the amount of such content, fansubbing communities grew. Although some online translation has become monetized (see Zhang 2016, 65), fansubbing is largely undertaken by distinct online communities, whose members engage in such volunteer immaterial labor for purposes of learning, sharing, and pleasure/entertainment (Hu 2012; Zhang 2016).

Feminist fansubbing embodies these same principles with the added purpose of combining enjoyment with advocacy that is not necessarily aimed at policy shifts, but rather raising awareness and changing hearts and minds. The object of fansubbing undertaken by some of the young women in this chapter certainly had content that more closely aligned with ideas associated with *nüxingzhuyi* feminism. For example, one participant who dabbled in fansubbing loved the US television show *2 Broke Girls*, which presented a version of popular feminism in its portrayal of young women's individual empowerment, often through consumption and sexual agency.[72] Other subject matter, however, was chosen to present strong messages regarding structural gender inequality and, like *OVO*, to bring to light taboo topics such as sexual assault.[73] Some content was humorous while some was extremely painful, but either way the purpose was to move the viewer.

One fansub example sent to me by a member of a feminist fansub group, illustrating how humor can be used to deliver a powerful message, was a three-minute video clip from *The Ellen DeGeneres Show*, the long-running daytime US television show hosted by the American comedian Ellen DeGeneres. In the three-minute clip, Ellen makes fun of "lady products," such as a "gentle laxative" and a "Super Cute Emergency Escape Hammer" that are the color pink and/or have pink packaging and, not incidentally, cost more than their "non-lady" versions (see figure 4.4).[74] Ellen calls these products "ridiculous" and ends the segment by stating, "We want the same thing for the same price . . . and a changing room with decent lighting. . . . And a *Friends* reunion," all to increasing audience applause. Although openly lesbian, Ellen is not known for taking firm stances on queer or feminist issues on her show. By equating not wanting products that price-gouge women with desiring "changing rooms with decent lighting," which is presumably a stereotypically women's concern, and a *Friends* reunion, she uses humor to soften her message. What had the power to be a strong critique of sexism as it manifests in the marketplace then became a joke, made for the sake of audience ratings.

In the feminist fansub group's hands, the *Ellen* clip was repurposed and uploaded to a video-sharing platform. Although the Chinese subtitles they added straightforwardly conveyed the original meaning (there were English subtitles as well), they put forward a much clearer feminist message through modifying the title and taking advantage of the space to add a description below the video. The clip posted on "TheEllenShow" official YouTube channel is entitled, "Ellen Reviews Products Made just for Women," and the accompanying text reads, "Ellen noticed that stores have regular products, but they also have lady products— which are the same thing, only pink—and she reviewed a few of them on the show." The feminist fansub video clip is named, "Ellen Vigorously Ridicules Pink Tax," and the much lengthier description explains gender-based price discrimination and includes a statistic that over 40 percent of women's products cost more than those for men (although no source for this figure is provided). Using language such as "vigorously ridicules" in the title, and calling the increased price a "tax," a term Ellen didn't use, can be understood as an ordinary ethical act; that is, educating viewers about how corporations profit from reinforcing gender

Figure 4.4. A screenshot of the fansubbed *The Ellen DeGeneres Show*.

stereotypes. The young feminists put the clip online hoping it would leave an impression by harnessing laughter and pleasure but also by generating anger at "ridiculous" sexism.[75] Their fansubbing thus became a networked feminist killjoy practice that potentially mobilized emotions to create affective value. Whether such fansubbing could translate into civic action in China is still unlikely (Zhang 2016). Nonetheless, the fansubbing labor was just one aspect of the group's feminist praxis, one that was enjoyable and meaningful.

Video Production

The livestreaming and fansubbing just discussed were projects by young feminists who created content to share ideas about feminism and to *move* others in some way. These then became part of a networked assemblage of multiple forms of feminist activism and advocacy that were undertaken relatively under the radar. Part of this assemblage is also made up of a range of content created by feminists who work or volunteer for NGOs. Here I discuss one young feminist's ideas about the purpose of the videos she created while working for a feminist NGO. Ting's involvement in the NGO grew out of her exposure to feminism in college, where, during a relatively freer period, she was able to attend academic lectures on feminism and listen to invited speakers from NGOs that advocated for sex workers and LGBTQ individuals. The NGO where Ting worked generated news reports about gender issues, developed educational materials and conducted training programs on topics related to gender and sexuality, and engaged in advocacy for marginalized groups. Ting's job consisted mostly of producing content to post on the NGO's channels on various social media platforms.

As mentioned earlier, Ting identified as a *nüquanzhuyizhe*, and, like Fang Ai, did not shy away from a public declaration of righteous anger on her WeChat public account. For example, the "about" section, which gave a brief introduction to the account, included a Chinese word that could be interpreted as "toxic" in it. The account name ironically appropriated the sexist tropes about feminists being toxic or angry women—-i.e., the feminist prostitute or killjoy—as an ironic stance. However, the name also evoked moral outrage and passion that moves one to act to right perceived wrongs.

In contrast to the evocation of something negative, the videos produced by the account drew more on humor and fun to communicate their messages. The content was very entertaining while also giving voice to the unspeakable. For example, one video was devoted to explaining gynecological health. Talking openly about something considered sensitive or taboo was also an effort to remove any stigma associated with it. Other videos addressed more emotionally painful topics, such as the extreme pressure to conform to dominant beauty norms and sexual harassment. Speaking of the videos, Ting said:

> They are all pretty personal women's experiences interpreted in this feminist way. Women are always anxious about how they look and [we discuss] why this is happening from personal experience and [explain] why we should stop doing this because it's all the societal standards. . . . We shouldn't be anxious about that because we are all ordinary women. We should resist that, the consumerism and things like that.[76]

Another video took on the "leftover woman" discourse, mentioned earlier, and featured two women, one of whom was celebrating her 27th birthday. She pretended to cry, which led to a discussion about feminism. The young woman says her friends ask her how they could become feminists and how one could know if they are a feminist. Her answer was straightforward: "If you support gender equality, and if you think you are a feminist, you are!" The women then discuss why feminism is "fun." With humor and laughter, they next assured their viewers that "there are no perfect 100% feminists in the world, and nobody needs approval from someone else to be recognized as a feminist. It is one's own self recognition." The message was light and uplifting, meant to take the pressure and anxiety away from the serious concerns of feminism, at least momentarily, and attract people with a very open definition of feminism.

When asked if the videos she produced received negative feedback, Ting said with a laugh, "No. It's kind of weird. We should have haters, then we would get more popular." However, popularity was not the point; rather, drawing others into the networked feminist killjoy assemblage was. Ting also made livestreams that were similarly lighthearted. However, eventually her Bilibili livestreaming channel, like Fang Ai's, was shuttered.

* * *

These examples of live streaming, fansubbing, and video production represent a minute fraction of the feminist content my informants created. They emerged from an ethical impulse to ultimately change society, and all were designed to stir others by addressing a range of issues. The content tapped into pain, frustration, and a refusal to be silenced, and women shared their experiences using humor and channeling anger to educate and create community, all of which have been documented as hallmarks of networked feminism (Clark-Parsons 2022; Ging and Siapera 2019; Mendes, Ringrose, and Keller 2019; Rentschler and Thrift 2015). Other hallmarks, unfortunately, include anxiety, exhaustion, and burnout.

Self-Presentation and Self-Preservation

Sometimes I think if I weren't a feminist, I would be happier.
—Fang Ai, born in 1996

The women featured in this chapter were proud to be feminists. It brought them meaning, connection, and pleasure, gesturing to the *joy* that came with being perceived as a feminist killjoy. However, this ethical commitment that began with the self and moved outward could bring intense pressure to *feel* like one was making a difference and to *perform* a feminist identity, often via social media. When a young feminist's efforts were met with scorn or vitriol, obviously this brought further emotional stress. In this section, I document pressures related to self-presentation and concerns about efficacy as well as anxiety and fear caused by more serious harassment and threats. I then briefly discuss the women's modes of self-preservation and self-care/restoration.

Affective Labor and Feminist Self-Presentation

All the young feminists mentioned various pressures they felt to create a certain feminist identity online. On the one hand, anyone who has a presence on social media has most likely at some point experienced at least mild anxiety when determining how to display a version of the self.[77] As I have detailed in previous chapters, domestic workers strove to present a positive and morally upright online identity to counter discourses about their backwardness, while young creatives used social media to cultivate their personal aesthetics. The impression management demands that I document here are qualitatively different, in that the feminist activists had to attend not only to the ordinary pressures that most young adults experience online (see Liu 2011), but also those that resulted precisely from their feminist passions and the publicness of their commitments. Like other activists, their work required a large amount of time and affective labor (Gould 2009; Rodgers 2010), amplified even further in the era of social media (Mendes, Ringrose, and Keller 2019).

Pressure, Disappointment, Discipline, and Violence

Most of the young feminists had to manage their presence in numerous online spaces, such as on Weibo and through participating in several WeChat groups related to feminism and/or queer issues. Those who administered the groups and/or were more well-known felt a strong need to share helpful information and to post frequently about news related to gender and/or sexuality. For example, regarding her WeChat group, Jia said, "The purpose is to help [the members] feel closer to feminism. . . . I hope feminist ideas will expand to more and more organizations. We share these ideas and information and help each other."[78] Jia implied that moderating the group could be stressful, but other participants were more direct in describing the pressure they felt to be up to date on the latest gender justice issues or risk being perceived as disengaged. This kind of pressure was especially prevalent when using WeChat, where a user's friends' circle (Moments) is made up of strong and weak ties, unlike Weibo, which offers more anonymity. Those whose work was connected to feminist issues felt even more stress because of their higher profile across platforms (which meant more scrutiny) but also due to their sense of duty. As one related:

> Especially when it's a time like International Women's Day or the Women's March or . . . some event just happened, like #MeToo, and we have another university [where] some professor harassed his student, people keep posting news. So, I'll keep posting because I want my friends [who don't follow] these issues to know this stuff.[79]

She then added that she was tired of keeping up and joked about posting "something politically incorrect," such as "I'm sick and tired of seeing all the people just posting the same thing over and over." She did not feel like she could do that, however, stating, "Yeah. . . . Maybe people won't like me. It's a kind of peer pressure from the feminist community," and then laughed.

This statement, although said as a joke, pointed to another source of stress for many participants: the fear of being in an echo chamber. Within this networked feminist killjoy assemblage, many online groups had several overlapping members. As a member of several groups myself, I have witnessed how the same information is frequently shared across groups and then commented upon by the same people in those groups. The statement above continued:

> I used to browse Moments all the time, but now I've stopped doing that. It's kind of a waste of time. . . . It's sort of boring because you all have the same friends with the same values, thoughts, and interests, so basically like. . . . All my friends are posting gender news. . . . Because I produce that, sometimes I kind of stress out. Maybe it's, like, irrelevant. . . .

I will return to the comment about it being "boring" later, but here I want to highlight concerns about relevance. This problem is widespread among numerous forms of digitally mediated activism. These feminist activists' goal was to fight patriarchal structures, but if they were all talking only to each other, what did that mean for the efficacy of their tireless efforts?

A related concern was not getting the desired result from a post, more specifically not gaining the hoped-for response and instead being met with silence or being silenced by family members and friends. One example mentioned by several women that illustrates both silence/being silenced occurred after the murder of a young woman by a Didi (akin to Uber) driver. This incident was a trending topic on Weibo, with many posts saying the young woman should have taken precautions to stay safe. Faye, a graduate student, was angered by the sexual double standard that held the young woman responsible rather than the driver. She also thought Didi was not doing enough to ensure their drivers were not predators. She posted a lot about this news but acknowledged that "my friends who commented . . . are basically the people who agree with me, who also feel angry about the issue."[80] Her family members had no reaction, however, which caused her to feel "very disappointed." Another participant posted a similar sentiment as Faye's and said that later her mother asked her why she was posting so much negative news and told her she should uphold a more positive image. Such silence and discipling by family members accords with notions of maintaining face, in this case by not causing one's mother to lose face as a result of her daughter's "negative" posts. It also reveals how the state discourse of "positive energy" works as a governing technology, circulating online and internalized, thereby rendering only certain posts acceptable for preserving "civility" in neo/non-liberal China.

The disciplining by family members and sometimes by friends for calling attention to the gendered dynamics of "negative" stories was an irritant to be sure, but not a deterrent. The irritation brought by these experiences was mild compared to the anxiety caused by posts that were censored, accounts that were deleted, and the labor required to regather contacts, reboot groups, etc. Some participants also suffered very violent harassment and trolling. Just as Fang Ai shuttered her livestream out of fear, other participants relayed stories of facing increasing hostility online after the Golden age of Weibo ended. Censorship increased, more restrictions were promulgated by the state, and violent nationalists, often supported by the government, went after anyone who did not conform to the status quo, particularly feminists. As Ning stated, "Since 2015 or 2016, Weibo has become scary because a lot of people curse (*ma* 骂) and attack (*gongji* 攻击) you."[81] Both high profile and relatively unknown feminists have continued to be the targets of violent threats and harassment. In some ways, their experiences parallel that of feminists across the globe who, as a result of their involvement in offline campaigns and online hashtag activism, are subject to violence and misogyny (Faniyi 2023; Gleeson 2015; Jackson, Bailey, and Welles 2020; Mendes, Ringrose, and Keller 2019). Of course, feminists can also disagree among themselves in ways that cause fallout and pain. All of these experiences could lead to emotional and physical stress and burnout.

Boredom, Burnout, and Self-Care

Boredom, exhaustion, disappointment, frustration, and outright fear can accompany the joy and exhilaration of being an activist in diverse contexts, and this was no less true for the young women featured in this chapter. The "intensities of emotion" that come with work that seeks to remedy injustice also can lead to burnout and fatigue (Gould 2009), which in turn necessitates self-care and restoration.

Some strategies for dealing with these consequences of feminist activism are already built into the digital platforms. For example, WeChat offers a variety of Moments settings which give users control over who views their content.[82] WeChat also allows members of groups to mute group alerts. Some participants, who were members of anywhere from 10 to 20 WeChat groups related to feminism and/or queer issues (including anti-domestic violence, anti-sexual harassment, and women's health), muted all the alerts. One participant mentioned a feminist activist who did not use the mute function and regularly had more than 100,000 notifications (!). It is important to note that, in addition to WeChat and Weibo, many participants were active on other social media platforms, including Douyin (TikTok) and Xiaohongshu (Little Red Book). Thus, the sheer volume of social media activity—sharing content, creating content, reading, commenting, and the like—could become overwhelming.

An individual strategy of some women, not surprisingly, was to check out, at least for a while. For example, Jia discussed how as a high school student, she shared a lot about feminism and got into online debates with her classmates. Even though these debates were at times "difficult," she enjoyed engaging others. As a university student, however, she said, "I rarely see something I want to forward. It makes me tired. . . . Because of my classes and work, I'm really lazy now."[83] Jia's experience of being tired is understandable, yet her critique of herself as "lazy," a word she used two other times during this part of our interview, revealed a lack of self-compassion or recognition of the battle fatigue she experienced and that she needed to take a break.

Other women did recognize the need for self-care, which instead of unplugging meant *engaging* with social media. After one participant expressed her concern about being in an echo chamber and being burnt out, she discussed her love of Weibo, which, as mentioned earlier, allows for more anonymity. She stated, "Weibo is our personal paradise . . . to relieve pressures, like maybe we follow stars, and trace their gossip, like normal girls do (*laughs*). . . . (O)n Weibo, we are just like normal people. We just enjoy, like normal, boring stuff."[84] Although activism can be tedious, here boredom was what was desired. It can be an "ugly feeling," yet it also comes with no expectations from others, only enjoyment for oneself. Other young feminists mentioned escaping into boys love *manga* for similar reasons, again as a focus on the inner self as opposed to the outward realm.

All the women expressed the need at times to take a step back and regroup and care for their selves, even if they did not experience online violence or harassment. However, none remotely hinted at giving up on their feminist commitments, even as some acknowledged that they didn't know if they could ever change entrenched gender inequality.

Conclusion

The presence of feminism in China has always generated contestation between those who have passionately promoted feminism as a means of social, cultural, and individual transformation, and opponents who have called feminism inappropriate for China because feminist ideology is supposedly anti-male, anti-family, and not suitable to China's unique history and culture. In neo/non-liberal China, institutionalized sexism has remained rampant, while Xi Jinping has been quite overt about his desire to reinvigorate binary gender roles, partly because these are associated with so-called traditional Chinese, as opposed to western, virtues. Also, as China faces a demographic crisis, it is a means of encouraging women to stay in the home and have children. Simultaneously, spaces for contestation have shrunk in the public sphere. Yet, feminism has nonetheless been embraced by young women—mostly urban and educated—as a means of sense-making and working toward societal transformation, and as part of a project of constructing an ethical self, even as this identity brings the very real risk of social, economic, and/or political marginalization.

Contemporary "Chinese feminism" is sometimes presented as a singular resistance to a patriarchal state by a unified group of closely-allied activists using social media rather than as a multi-faceted and evolving formation.[85] In this chapter, I have documented how young women from different walks of life wrestle with the meaning of feminism, and how a range of affects, feelings, and emotions—anger, sadness, fear, happiness, and joy—animate their feminist actions both large and small. This chapter has included the voices of some seasoned activists, but I have also tried to highlight the work of "ordinary," unknown young women whose passions move them to create, educate, and use social media to have a voice to express their views on sexism, patriarchy, misogyny, and gender violence. This focus is not because I believe that some force—be it marketization, the rise of the internet, or a social movement such as feminism—is going to transform China in the direction of political liberalization. On the contrary, the Chinese state and its propensity for "authoritarian resilience" (Nathan 2003) has been remarkably consistent in dashing these hopes. Rather, I have sought to show how young feminists work toward small social transformations.

The networked feminist killjoy assemblage I have documented in this chapter has multiple articulations as it continues to be crosscut by transnational cultural and intellectual flows, generates contested meanings and namings, and uses multiple methods to achieve diverse goals. My informants nonetheless saw themselves in common cause with one another, that cause being to dismantle patriarchy in China. In doing so, they are nodes in a networked feminist killjoy assemblage. Other nodes in this assemblage, not featured in this chapter, are more radical. The decentralized nature of this assemblage is crucial in a sociocultural formation like neo/non-liberal China, where feminist voices that advocate for social change are most often not valued, at least not in Couldry's (2010) sense of valuing voice, which means making space for and respecting a range of narratives. Instead, feminists are often ridiculed, undermined, threatened, and/or censored by various forces, in particular the party-state, social media platforms, and many members of the general public. Sometimes this causes annoyance (e.g., when ridiculed), but other times it creates significant anxiety and burnout, especially for those who are devoted activists trying to change structural gender inequality and proudly identifying as *nüquanzhuyizhe*, regardless of the costs. In either case, there is the necessity of self-care. Such care can take the form of checking out momentarily, escape, or just wanting to be "normal girls" enjoying celebrity news, mindless entertainment, and the like.

Feminist actions have led to some social changes in China, but the longer Xi Jinping has held power, the less space there is for an emancipatory feminist politics to exist without censorship. Nonetheless, glimmers of feminist ideas have diffused into popular culture, not just the consumerist lifestyle feminism of certain social media influencers or in television shows that contain elements of popular feminism, but also in shows like *Imperfect Victim*, a 2023 web series that chronicles the investigation of a workplace assault.[86] It is hard to believe that this show would exist without China's #MeToo movement. Thus, I prefer to end on a hopeful note that the efforts of young feminists will slowly improve the lives of all Chinese women.

Conclusion

Affect, Ethics, and Marginalized Voices

In early March 2013, as I had recently begun extended fieldwork in Beijing, the first message I received on my newly created WeChat account
was from one of my closest Chinese friends, who was also my first
WeChat contact. It read, "North Korea tested a nuclear bomb!"[1] It was a
startling message, to be sure, but what is more surprising in hindsight is
how in the intervening years the geopolitics around North Korea have
barely shifted (at the time of this writing) while so much in China has
changed. After that first message, WeChat would soon evolve from a
mobile messaging app into a Twitter-meets-Facebook-meets-gaming-
meets-do everything-in-your-life social media platform that is basically
impossible to live without in China. That spring I used Instagram in
order to interact with a group of domestic workers involved in an NGO
project. A year later Instagram was blocked by the Chinese government,
and it would continue to be off and on until it, along with nearly all other
US-based social media platforms, were eventually permanently blocked.
Since that time, some formerly popular Chinese apps have disappeared
while still others featured in this book in addition to WeChat—Weibo,
Taobao, Bilibili, and Douban—have remained fixtures in the vibrant
world that is Chinese social media.

In early 2013 as well, Xi Jinping was at the beginning of his ascent to
power, and over the years he would become China's uncontested leader,
intent on realizing the "Chinese Dream" while in the process transforming economic, social, cultural, technological, and political realities. Along with marketization and privatization, the government has
exerted intense effort to shift China's economy from export-processing
to high-tech and innovation. The state-owned sector has also been bolstered, as both a disciplinary tactic to assert control and as a means
of promoting and safeguarding domestic innovation. All the while,

discourses regarding entrepreneurship, creativity, and innovation have continued to diffuse among all levels of society and have become articulated to the economization of social and cultural life.

Within this milieu, the authoritarian party-state has promoted "positive energy" and "Chinese values," the latter a mix of Core Socialist Values and a selective array of Confucian tenets, including patriarchal gender ideologies that locate women's primary role as that of caretaker of the home and family. Xi has firmly and resolutely asserted his power while invoking a therapeutic mode of governance that induces desires and feelings—for personal happiness and a fulfilling life—while channeling these toward state goals. Harnessing affect, stirring passions, and circulating particular emotions—pride, patriotism, and gratitude for China's leaders and contempt for and anger at any person or entity that expresses opposition and/or appears to be holding China back—are tactics meant to summon the populace in accord with Xi's Dream. However, the psychologization of society, whereby individuals hope to find happiness and fulfillment through an array of self-help, spiritual, and/or religious practices reveals that wide swaths of the population seek meaning in addition to or outside of state regulatory discourses. Informatization, platformization, and datafication are intertwined with these economic, political, and socio-cultural phenomena and have enabled the integration of social media into every aspect of life, as well as greater state censorship, repression, and surveillance.

This context, which took shape over the years I was conducting the research for this book, is what I call neo/non-liberal China (admittedly not the most elegant term). Theoretically, the concept of neo/non-liberal China is indebted to earlier formulations in the 2000s that drew on Foucault's notion of governmentality and the enterprising self, such as Ong and Zhang's (2008) "socialism from afar." These were meant to capture the interplay of newly emerging micro-freedoms, the unleashing of self-interest, and the expression and fulfillment of desires that previously were either hidden or outside the realm of discursive logic until unleashed by the state's embrace of marketization and privatization. The authoritarian party-state still set the boundaries for political expression but seemed to have taken a step back from its former overt intrusiveness into daily life. However, by the late 2000s and early 2010s, even before Xi Jinping's rise, this context seemed to be shifting. Over time, a more repressive state has decreased certain of those micro-freedoms to what could be called nano-freedoms. While self-interest still abounds, individuals must take greater care in how it is lived out. To be sure, Xi's project of national rejuvenation needs self-responsible, self-optimizing, entrepreneurial subjects, and the Chinese Dream is meant to beckon all. Clearly, people comply, ignore, resist, or respond in all manner of ways to this call, stirred by their own feelings, desires, and ethical judgments. Thus, the "non" in "neo/non-liberal" does *not* mean no freedom; rather, it captures the fact that outward expressions that do not conform to dominant orthodoxy are severely constrained or eliminated in ways in which they were not from the mid-1990s to the late 2000s. Neo/non-liberal China is meant to offer a tentative framework for capturing this historical conjuncture.

Across the chapters in this book, I have documented how, within this evolving assemblage, the aspirations, self-expression, self-representation, fights for equality, maintenance of community, and entrepreneurial practices of those who are socially, economically, and/or politically marginalized are articulated to social media. Through research among marginalized young creatives, rural micro-entrepreneurs, domestic workers, and young feminist women, I have argued for the importance of considering the affective and ethical dimensions of these processes. This focus is not meant to oppose the emotional to the rational; rather, it emerges from the socio-cultural context of China, where dominant understandings of personhood emphasize a holistic conception of mind/body that encompasses thought, emotion, affect (*gan*), energy, and the sensory. In this book I tried to show, in Jie Yang's (2021, 977) words, "the formation and circulation of subjects through affect" and also "the circulation of emotions between subjects." Similarly, in analyzing ordinary, quotidian practices, my intention was not to ignore what I have called "daunting" forces (AI, big data, massive surveillance techniques). On the contrary, companies such as Alibaba rely on these as they try to bring economic prosperity to the countryside, as discussed in chapter 2. Moreover, as I have stated many times, ramped-up surveillance, censorship, and repression are part and parcel of neo/non-liberal China. Some of the young creatives had to deal with anger and loss associated with being censored for their ideas, as did many young feminists, who were also the target of disembodied misogyny as online trolls harassed them. For many people featured in this book, however, as they tried to make their way and live their lives, oftentimes their main challenge was to negotiate long-term structural constraints that manifested in very material ways: a lack of infrastructure, a demanding and/or exploitative employer, patriarchal gender relations embodied in an unethical relative, a disapproving parent, and so on. In using social media to try to deal with these obstacles and pursue their goals, the outcomes could be contradictory or ambivalent.

The young creatives faced challenges due to their marginalized status as outsiders in Beijing and their lack of prestigious degrees. Of the four groups examined here, they were the most clearly linked to state discourses and policies regarding creativity and innovation yet also insistent on their embodied passions as compelling their movement to the city. Social media was articulated to such movement and to their learning, as well as to their constructions of taste and distinction and forms of curation, the latter of which could be tied to an ethical imperative to use their creativity for good in some, but not all, cases. Long-term success ultimately depended on their social and cultural capital, and whether and how these could lead to economic opportunities that were meaningful and not exploitative. Some of them, like millions of young creatives around the globe, are still searching for that dream.

The rural micro-entrepreneurs also aspired to better lives, but what that meant to them was explicitly tied to sustaining their livelihoods. They faced challenges due to lack of infrastructure, and due to relational ties and obligations. Highlighting how the entanglement of *guanxi, renqing,* and *ganqing* could enable or constrain their endeavors brings to the fore the affective and ethical dimensions of digital inequality. In particular, in the case of middle-aged rural women, who see their primary role as caregivers in the family, I have argued for the importance of understanding how they use technology in ways they deem appropriate, which is not the same as examining technology appropriation. Rural women's lives are governed by patriarchal gender norms, which lead to the devaluing of their labor, as do dominant discourses that construct them as having low quality, or *suzhi*. In managing both care work and micro-entrepreneurship, they deployed social media in a manner that fit their needs—whether with a sewing machine and a WeChat group, an attempt at a new but ultimately unsuccessful form of corporate e-commerce, or a platform business that existed in a gray area.

Domestic workers, most of whom were from rural areas, had to contend with being outsiders who were also viewed as having low *suzhi*. As they were relatively powerless in the urban home, they used social media in a way that enabled them to take control of their own representation and present a positive, caring, and knowledgeable subject. Their use of social media was an individual project of self-development and a highly relational means of supporting one another and maintaining their small community. Their positivity should be seen as a tactic to counter the negativity that is attached to them, and it is one facet of their communicative empowerment, which nonetheless does not give them a voice in the wider society.

Of all the groups featured in this book, the young feminists as a whole tended to have the most social, cultural, and in most cases, economic capital. Their passions moved them to try to transform patriarchal gender norms and bring about greater gender equality. As the heyday of on-the-ground feminist activism had passed after 2015, social media became the primary domain where they could work out the meaning of feminism—the terms to use, the goals to try to achieve, the messages to put out. As part of what I have called a networked feminist killjoy assemblage, the young women did not necessarily always agree with each other. Nonetheless, they were aligned through their ethical commitment to use a variety of means—blogs, livestreams, video production, and fansubbing, just to name a few—to spread knowledge and hope. The misogyny and hate some of them faced in the process meant that many had to engage in self-care and healing, which could involve totally unplugging or seeking out online spaces away from activism and that brought joy and pleasure.

Organizing each successive chapter around the individual, the family, the community, and the network was done for analytical purposes, yet clearly individuals are always embedded in families, communities, and networks. While the young creatives were unique in their explicit pursuit of personal aesthetics, all of the people in this book participated in some way in processes of self-creation and meaning-making that was articulated to social media. Similarly, many young creatives wanted to use their art for social good, just as domestic workers sought to uplift those in their small mediated community, and feminist activists endeavored to bring hope and healing to others. Many of the rural micro-entrepreneurs also saw establishing their businesses as a means of helping not only their families but also their fellow villagers. All the participants engaged in various types of affective labor, even if for different purposes. Although each chapter elaborated on different aspects of social media use, the affective and ethical dimensions of these pursuits have overlaps, as all the people in this book faced challenges and uncertainty in the shifting and evolving neo/non-liberal assemblage.

Still, given the realities of structural inequality, the urban/rural divide, gender norms, and generational differences, these four groups had diverse goals and had to contend with distinct challenges. In many regards, older rural women, both those who were micro-entrepreneurs and those who were domestic workers, had little in common with educated, middle-class, (mostly) urban-born young feminists. However, they were all disciplined by patriarchal gender norms and discourses regarding harmony and civility that are ultimately about subordination and control. Like the young feminists, the young creatives on the surface were also far removed from the older women in terms of social and cultural capital. However, like the older women, they too were from rural towns and villages and had rural *hukou*, and many struggled with economic insecurity and could be subject to exploitative labor conditions.

Linking all of these individuals and their aspirations to transcend circumstances and unpacking the affective and ethical underpinnings of these pursuits raises certain questions, namely, is there a risk of flattening the very real differences between them due to class-, age-, and gender-based hierarchies, geographic location, and material circumstances? Placing these disparate voices in metaphorical conversation with each other is meant to allow for the expression of Bakhtin's (1981) heteroglossia, which acknowledges stratification within the multiplicity of voices. It is also one way to materialize Nick Couldry's (2010) "voice as value," as all of the voices in this book are meant to be valued equally. Feminist activism and creative industries are hot topics in academic scholarship and popular culture, yet the struggles of domestic workers and rural micro-entrepreneurs, particularly older women, challenge us to leave aside our own biases and recognize their passions, ingenuity, and resilience, and that their voices should count when we consider larger social and cultural transformations taking place in contemporary China.

The processes and practices I have examined in this book, across diverse groups, places, platforms, and times, in many ways were magnified and multiplied by the COVID-19 pandemic. As subsequent variants continued to emerge, weeks- and even months-long lockdowns ensued in various locales, with daily testing and ramped-up surveillance through the health code app, all as part of the government's Zero-COVID policy. Thus, in China (as elsewhere), social media became the primary means for expressing deep feelings of grief, anger, anxiety, and frustration—at government coverups, lockdowns, deaths of loved ones, and so on. Social media also was a lifeline for dealing with seemingly endless boredom and despair; for bringing hope and optimism; and for crucial social support and connection, economic livelihood, well-being, and, stated with no exaggeration, survival.[2] At the same time, through the corporate-state health code app and other regulatory measures, social media was linked to intensified surveillance, control, and repression. Yet, the app also enabled people to return to their workplaces, though not without glitches, once the initial wave of COVID was thought to have been contained.

The pandemic also further amplified many of the ordinary affective and ethical processes I have highlighted in this book. As just two examples, domestic workers across China faced particularly challenging circumstances: many were forced to stay with urban employers under lockdown, with no reprieve and with no days off, or they were stuck back home in their villages with then no means of generating income. Their counterparts who had jobs as maintenance staff of residential compounds had to live for weeks in basements, at the mercy of donations of food and clothing from the residents of the compounds. Despite their pain, the private messages that a handful of domestic workers sent me during this time still reflected a positive outlook, however, as they sought to make sure I was doing okay during COVID. Feminists individually faced a range of experiences, but collectively, including some of the participants in this book, used hashtag activism to rail against the blatant gender discrimination revealed in the state response to COVID-19. They engaged in online protests against what appeared to be the forced shaving of the heads of some women who were frontline healthcare workers and against the state giving others birth control pills to delay their periods while refusing donations of sanitary napkins, among other things ("Outcry" 2020; J. Zhang 2023). Not surprisingly, some of these feminist protests were met with online misogyny and vitriol. These

were joined by extreme nationalism, which is another hallmark of neo/
non-liberal China. In Xi Jinping's China, directing anger towards feminists is a way to kill several birds with one stone—shoring up patriarchal
gender ideologies that seek to keep women from having a voice, silencing those who refuse to stay quiet, and averting attention away from
structural problems.

When protests finally broke out in late November 2022 over the seemingly never-ending lockdowns, of course social media rapidly spread the
news. Young women were often at the forefront of these on-the-ground
protests, just as they had been earlier in the year when two incidents generated national online outrage and protest: first, a viral video in January that
showed a woman in rural China chained in an outdoor shack, wearing no
jacket in the freezing cold. The overwhelming response online was shock
and dismay, with many wondering, how could this occur in modern-day
China? When it was revealed that the woman suffered from mental illness and had been trafficked and had born eight children to her husband,
the outrage only increased. This case and a brutal, unprovoked beating of
four young women at a late-night barbecue restaurant in Tangshan in June
added to the already extreme anger, horror, and despair that reached a tip-
ping point in the late fall protests.[3] By the time the Chinese government
abruptly dropped its draconian Zero-COVID policy in December 2022,
these emotions had only multiplied. If Xi Jinping desired to spread positive
energy, there was a wave of "negative energy" engulfing China.

Shortly after the Zero-COVID policy ended, a friend in Beijing sent
me frequent Twitter (now X) messages with secretly filmed videos of
coffins and makeshift morgues hidden behind plastic on the back streets
of Beijing. Then, only a few weeks later, another friend in China, referring to the pandemic and lockdowns, said, "It was almost like it never
happened." The government wanted to move on, which meant that deep
anger, grief, anxiety, and fatigue among wide swaths of the population
could not be openly expressed. This calculating mode of state-enforced
amnesia is reminiscent of China in the wake of the Tiananmen Square
massacre (Lim 2014). However, tightening ideological control by the
party-state cannot suppress people's intense emotions, including the
deep feelings of malaise wrought by the pandemic and magnified by
economic uncertainty post-lockdown. Such sentiments are particularly
prominent among younger segments of the population that are bearing

the brunt of the economic downturn and alarming youth unemployment rates, and especially among young women who are told to have more children to help contain a demographic crisis (Stevenson 2023).[4] Throughout 2022, the "lying flat" (*tangping* 躺平) phenomenon that had spread in 2021, discussed briefly in chapter 1, was followed by increasingly disconsolate neologisms, as seen in just two examples: "let it rot" (*bai lan* 摆烂) and "we are the last generation" (*women shi zuihou yidai* 我们是最后一代).[5] These sentiments, expressing hopelessness, despair, and nihilism, have raised concern among China's leaders, whose response thus far has basically been to tell young people to suck it up and contribute to their country (Ni 2022a).

But China, of course, is huge, and not all people suffered equally during the pandemic, nor have all lost hope. The rural residents featured in chapter 2 actually fared relatively well, as lockdowns were not as severe and prolonged. Rural revitalization and urbanization efforts continued to move forward in their village and countless others, bringing natural gas and running water inside people's homes, which will soon also have flush toilets, the result of the government's "toilet revolution."[6] With these new luxuries, and as pandemic fatigue and discontent with the stresses of city life has gripped much of China, my rural host and her family started to spend more and more time in the village, and she told me she hopes to retire there. Will these external changes transform the patriarchal gender relations that seep into every aspect of village life? Most likely, the answer is no. But better roads, faster Wi-Fi, and more comfortable homes will surely be a draw to young people and migrant workers, exhausted and sick of the treatment they continue to receive in cities. Discontent on the part of young urban white-collar workers with the 996 lifestyle (9:00am–9:00pm, six days a week) has made rural areas more attractive to them as well (*before* they retire). Still, transforming the countryside so that it is a place where rural young people can realize their ambitions is a daunting task.

All of these phenomena confirm the importance of understanding the role of affect and emotion and the construction of the ethical self in daily life practices that are often in some way mediated. As Lawrence Grossberg (1992, 80–81) argued long ago, affect "operates across all of our senses and experiences, across all of the domains of effects which construct daily life. Affect is what gives 'color,' 'tone' or 'texture' to the lived." He adds that affective relations potentially are "the condition of possibility for the optimism, invigoration and passion which are necessary for any struggle to change the world." Most of the participants in this book could not necessarily be said to be trying to change the world, though some hoped to change Chinese society. Through their daily struggles and pursuits, in which ordinary ethical choices had to be made, all engaged in individual transformations. This final discussion reveals that this book is about much more than Chinese social media; rather, my focus on the assemblage of social media, bodies, emotions, embodied passions, ethical choices, objects, and struggles for voice enables insight not only into individual experiences but also China's larger structural transformations. Social media is contradictory, unpredictable, and never neutral. It can be linked to depression, deception, exhaustion, exploitation, and violence. Within certain configurations, it is also part of a circuit for affective flows that can attune bodies toward the hope of a better future.

ACKNOWLEDGMENTS

This book represents over a decade's worth of research and writing that coincided with several, at times devastating, personal challenges in my life. In addition, much of the writing occurred during a years-long global pandemic. As a project that is grounded in the inarticulable and unpredictable affective forces that move bodies, it seems appropriate that this book has taken many unexpected twists and turns during its development. One consistency, however, has been the unfailing support, encouragement, and generosity of so many family members, friends, colleagues, NGO staff, and students in the United States, China, and other locales.

The research could not have been conducted without funding from the Department of Communication and Journalism, the Melbern G. Glasscock Center for Humanities Research, the College of Liberal Arts Ray A. Rothrock '77 Fellowship, and the Program for the Enhancement of Scholarly and Creative Activities, all at Texas A&M University. I am forever grateful to my wonderful friends and colleagues in the Department of Communication and Journalism at A&M, who were extremely supportive in numerous ways throughout the process of bringing this book to fruition: Kevin Barge, Hart Blanton, Sandra Braman, Patrick Burkart, Angelique Gammon, Jess Havens, Antonio La Pastina, Kristan Poirot, Srivi Ramasubramanian (now at Syracuse), Nancy Street, Rick Street, Lu Tang, and Anna Wiederhold Wolfe. A special thank you to Heidi Campbell for the numerous small gifts, meals, and words of encouragement that always lifted my spirits.

In Beijing, I am thankful to Zhang Yanqiu, professor at the Communication University of China, who hosted me as a visiting scholar. At CUC, Ji Deqiang, Dianlian Huang, and Steven Xu never tired of talking with me about social media and Chinese society, introducing me to people, and inviting me on outings. A big thanks as well to Geng Yiqun and Nancy Wang. At Tsinghua University, much appreciation goes to Guo Yuhua and Lu Jia for facilitating introductions with research participants. I can't express my profound gratitude to the different migrant worker organizations in Beijing that invited me to be part of their communities. I would never have met so many wonderful women, who taught me so much about perseverance, humility, and how to make the best out of any circumstances, had it not been for Yan Chengmei and Chen Jiyan, whose dedication to improving the lives of domestic workers is truly inspiring. I also thank Ying-Ying Lu and Qi Lixia, as well as Chen Shanshan, who always brightened my time in Beijing with her smile and support. Many thanks also to Stanley Chen, William Chen, Congshan He, Quan Quan, and Candy Wang for research assistance.

I always benefitted from the warmth and generosity extended by many friends when I was in Beijing: Bu Wei, Cai Yiping, Guo Yuhua, Liu Xiaohong, Liu Yanbin, Sun Wusan, and Xu Yang. Sun Wusan first introduced me to villagers in Shandong province and accompanied me on return trips. I am thankful to He Zhenbo for opening her family's home to us and hosting us with such generosity. Much gratitude also goes to Zhang Qi for accompanying me on my last trip there. In Beijing, I have always appreciated conversations with Ross Warner and his willingness to do whatever he could to facilitate my research. I am indebted as well to Shetou and Yuan Cai for stimulating conversations and introductions. Helen Zhang is always such a cheerful and affirming presence. I look forward to hanging out during many return visits. Thank you also to Angela Xiao Wu, Yu Sun, and Cui Xi, who, though not in Beijing, introduced me to people there.

I gained incredible insights from so many people who generously read and provided thoughtful feedback on chapters: Megan Ankerson, Sarah Banet-Weiser, Sandra Braman, Cai Yiping, Jenny Chio, Donnalee Dox, Mei-ling Ellerman, Joseph Jewell, Silvia Lindtner, Danchen Liu, Lü Pin, Elisa Oreglia, Nancy Plankey-Videla, Shen Rui, Joan Wolf, Fan Yang, Jie Yang, and Huiran Yi. Extra big thanks to Haoming Zhou, who read every chapter of the final draft of the manuscript. Much appreciation goes to Cathy Hannabach and Shazia Iftkhar, who helped me clarify my argument in each chapter. The manuscript also benefitted from conversations over the years with Arianne Gaetano, Rongbin Han, Min

Jiang, Bingchun Meng, Ceclia Milwertz, Jack Linchuan Qiu, Marina Svensson, Guobin Yang, Elaine Yuan, and Weiyu Zhang.

I am very grateful to many graduate students who joined me for parts of this journey. Watching them grow as scholars and learning from them was always inspiring: my current and former advisees James Cho, Marisa Doshi, Macy Dunklin, Jess Gantt-Shafer, Paige Jennings, Shelby Landmark, Danchen Liu, Caitlin Miles, Yongrong Shen, and Haoming Zhou, and students whose committees I served on: Brian Altenhofen, Wendi Bellar, Carrie Murawski, Kate Siegfried, Ruth Tsuria, Aya Yadlin Segal, Siyuan Yin, Felicia York, and Wenxue Zou. I also thank Shaohai Jiang for research assistance while he was a graduate student at Texas A&M.

I was fortunate to be invited to present portions of this work over the years at several institutions due to the generosity of the following people (some of whom have moved elsewhere since): Pauline Cheong, Center for Asian Research, Arizona State University; Eileen Chow (Duke University), convener of the 2015 "New Voices and New Approaches in China" Gender Studies Workshop, Fairbank Center for Chinese Studies, Harvard University; Josh Kun, Annenberg School for Communication and Journalism, University of Southern California; Sun Sun Lim, Asia Research Institute, National University of Singapore; Mirca Madianou, Department of Media, Communications, and Cultural Studies, Goldsmiths University; Jack Linchuan Qiu, School of Journalism and Communication, Chinese University of Hong Kong; Marina Svensson, Center for East and Southeast Asian Studies, Lund University; and Tianyang Zhou, at the UK China Media and Cultural Studies Association Biannual Conference, University of Leicester.

So many women friends and scholars have encouraged and sustained me along the way: Sarah Banet-Weiser, Anne Balsamo, Cai Yiping, Radha Hegde, Silvia Lindtner, Mirca Madianou, Barbara Sharf, Shen Rui, Joan Wolf, and Fan Yang. I am so profoundly inspired by and grateful to Jie Yang, whose work on affect and therapeutic governance in China has been a source of deep inspiration and who has been a truly generous and gracious colleague.

Wrapping up this manuscript coincided with a major transition in my life—moving to Ann Arbor to take a position in the Department of Communication and Media at the University of Michigan. I am so thankful to those in and outside the department—Megan Ankerson, Scott Campbell, Susan Douglas, Nicole Ellison, Hollis Griffin, Silvia Lindtner, Lisa Nakamura, Devon Powers, Dave Reid, Christian Sandvig, and Apryl Williams—who welcomed me and invited me into their homes for meals and other gatherings. I am forever grateful to Megan and to Susan for being sounding boards as I put the final touches on the book. I was fortunate to join the department at the same time as Pranav Malhotra, who is a brilliant scholar (and has the best snacks).

In addition to the many people already mentioned, I am so appreciative of dear friends in Bryan/College Station: Courtney Schumacher (you're a rockstar!), Jess Havens (love you, girlie), Mel Alvarado, Joel Griffin, Marisa Kaye, Tea Luck, Pascale Parker, Courtney Starrett, Kim Topp, Elise Winchester, Joan Wolf, and my former bandmates: Lauren Dewey Furhmann, Jonathan Hudson, Jason Parker, and Jeff Winking. Thank you also to Robin Bedenbaugh who long ago left Texas. It is also amazing to me that after leaving California many years ago, I still have wonderful friends who support me from afar and who never fail to make time in their schedules for me when I am there: Lucienne Aarsen, Gregory Anderson, Reka Clausen, Nancy Currey, Melina Dorian, Janet Goodwin, Linda Jensen, Gaby Solomon, Molly Smith-Olsson, Jen Tisdale, Tessy Tsoytsoyrakos, and Zulema Valdez (simpatico!). Judy Marasco, what would I do without you? Linda Rhine (I know you aren't in California anymore) the same.

I am grateful to my parents, Gene and Martha Wallis, and my sisters Inger Budke and Laura Wallis, who have always supported and believed in me.

I will be forever indebted to the many people in China who participated in this research and gave of their time and shared their experiences and insights with me over the years, not only about social media, but also about care, community, creativity, feminism, labor, and life, among many other topics. This book would not have been possible without their generosity and graciousness, for which I am so thankful.

At NYU Press, Eric Zinner has been extremely patient and supportive as I have brought this project to fruition. A huge thanks also to Furqan Sayeed, to three anonymous reviewers, and to the Critical Cultural Communication series editors Jonathan Gray, Aswin Punathambekar, and Adrienne Shaw.

Portions of the manuscript have appeared in different form in the following: "Micro-entrepreneurship, New Media Technologies, and the Reproduction and Reconfiguration of Gender in Rural China," *Chinese Journal of Communication* 8, no. 1 (2015): 42–58; "Domestic Workers and the Affective Dimensions of Communicative Empowerment," *Communication, Culture & Critique* 11, no. 2 (2018): 213–230; and "Social Media and the Ordinary: Marginalized Voices in Neo/non-liberal China," *Communication and the Public* 9, no. 4 (2024): 410–420 and are reprinted here with permission from the publishers.

NOTES

INTRODUCTION

1 In *Being Digital* (1996), Negroponte laid out his ideas about technology and how it would shape the future. At the "Big Talk," Negroponte gave a second keynote on "big ideas." One "big idea" was viral telecom enabled by satellite that would be affordable and accessible for all, apparently something he's been talking about since at least 2002: http://archive.wired.com.

2 The creation of the epoch-changing phone would entail tweaking the chip and creating a better battery that lasts longer and charges quickly.

3 The actors, only one of whom had been a migrant worker, included a former police officer and a college graduate. The play was co-written by the leader of an NGO that helps migrant workers.

4 The film footage showed old British factories and the inside of a library that contained documents about Karl Marx and labor in England.

5 For a comparison of the "new workers" with the older generation of industrial workers, see Wang and Qiu (2015). For more on the New Worker Art Troupe, see Yin (2020).

6 The ticket cost 100 *yuan*, definitely out of reach for the majority of migrant workers. Most likely those in the audience were given tickets by NGOs, as was the case for the domestic workers I knew that attended.

7 Tencent created QQ and WeChat. Alibaba is China's e-commerce giant and owns Taobao and Tmall. Baidu, Alibaba, and Tencent, often lumped together as "BATs," are three of China's most powerful and influential tech companies. In China, their founders are akin to Bill Gates and Steve Jobs.

8 Of course, the rise of domestic technology giants, such as Sina, Tencent, and Alibaba, was not only the result of ambitious and visionary innovators, but also strategic government policies and investment in telecommunications, seen for years as one of the primary drivers of economic growth and restructuring. A decades-long government-led informatization drive had resulted in tremendous development and diffusion of broadband internet, mobile phones, and smartphones by the early 2010s (Harwit 2008; Hong 2017; Wallis 2013a). After the 2008 global financial crisis, the state viewed telecommunications and high-tech industries as key to its desired shift to a consumer-based (rather than export-based) economy, as an innovator rather than manufacturer, and as a global leader in ICT industries (Hong 2017). Premier Li Keqiang's focus on "mass innovation"

and "mass entrepreneurship" (see Li 2015), which later became state policy, was accompanied by the government's Internet Plus plan, which was announced by Li when he gave his Government Work Report at the "Two Sessions" meeting in March 2015 (the "Two Sessions" refers to the annual meetings of the National People's Congress of the People's Republic of China and the National Committee of the Chinese People's Consultative Conference). Internet Plus seeks to "integrate mobile Internet, cloud computing, big data and the Internet of Things with modern manufacturing" and further develop e-commerce and e-banking (State Council 2015d; see also Hong 2017). Made in China 2025 is a comprehensive policy to enable China to become a manufacturing superpower and outlines the development of several strategic industries, including robotics, biomedicine, and aerospace (State Council 2015a). It emphasizes self-reliance and domestic manufacturing in these efforts. For discussions and assessments of Made in China 2025, see Erdenebileg and Hu (2017) and Zenglein and Holzmann (2019).

9 Silvia Lindtner (2020) discusses in great detail how making was attached to what she calls an "affect of intervention," in which modernist ideas about linear technological development and its equivalence with progress were joined to futuristic visions of life made better by technology, innovation, and creation. Entangled in this affect of intervention were socialist ideals of change and social justice along with neoliberalism's insistent imperative "that one convert the self into human capital," which necessitates "investing in various aspects of one's own life in order to make the self attractive to the machineries of finance and speculation and investment" (13).

10 The alleged reason for the suspension was Jack Ma's sharp public critique of China's regulatory system, but the crackdown was a result of a number of factors, including the government's desire to reign in big tech, which was seen as too freewheeling and too powerful; allegations of monopolistic activity and antitrust practices; concerns about data privacy and security as well as financial risk; and the government's response to public discontent over the tech industry's treatment of its workers in the context of growing social inequality (Chang and Goldkorn 2021; Cohen 2021). For a helpful timeline of the initial crackdown as it expanded across the tech industry and to seemingly unrelated domains, see https://thechina project.com. After numerous big tech CEOs stepped down, huge fines were imposed, especially to Ant, and the crackdown seemed to be winding down by mid-2023 (Huang, Zhang, and Zheng 2023).

11 Outside China, before COVID-19, perhaps the most well-known of these protests were the worker suicides at Foxconn in the early to mid-2010s that generated international outrage. In November 2022, workers protested and clashed with police at a Foxconn plant in Zhengzhou due to workers' concerns about safety and pay (McDonald and Soo 2022).

12 See Wallis and Shen (2018) for an early articulation of this concept.

13 Couldry (2010) elaborates that to value voice means to focus attention on "the conditions under which voice as a process is effective" as well as what forces,

organizational forms, and circumstances (subtle or not) work to "undermine or devalue voice as a process" (2). For an overview of various conceptualizations of voice, see Weidman (2014). I thank Jie Yang for this reference and for urging me to state my conceptualization of voice early on in this chapter.

14 My understanding of technology as articulation and assemblage is greatly influenced by the work of Jennifer Daryl Slack and J. Macgregor Wise (2015) (see also Wise 2005). They urge us not to focus on the "thingness" of technology, but rather to ascertain how technologies are one of multiple heterogenous elements that are articulated to form an assemblage, which can shift as it is deterritorialized and then possibly reterritorialized. For this thinking, they are indebted to Deleuze and Guattari (1987). For a more in-depth discussion, see Wallis (2013a, 13–15).

15 In the 2010s, in addition to Sina, several Chinese tech companies, including Sohu, Netease, and Tencent, had microblogging (*weibo*) platforms, but by the end of the decade, the *weibo* of the latter three had ceased operating.

16 On July 23, 2011, the train crash occurred in Wenzhou in the southern coastal province of Zhejiang. Later known as the "7.23" incident, the crash garnered significant public attention and outrage, both because it served as one more example of the human cost of China's rapid development and because the government lied about the number of casualties and tried to hide other details, even to the point of burying one of the derailed train cars (Bondes and Schucher 2014). The crash was later determined to be the result of both technical and human error. See "China Bullet Train Crash" (2011).

17 This point was driven home a couple of weeks after the Wenzhou train accident, when I visited a number of villages in Shandong province. Not far from Qingdao, some villagers knew of the government's version of the train crash via Weibo, and thus the fact that I had just taken a high-speed train from Beijing to Qingdao was an instant topic of conversation. However, the further I traveled inland, the less likely it was that any of the villagers (some of whom are featured in chapter 2) had heard of Weibo. Instead, middle-aged (mostly male) micro-entrepreneurs were using Tencent's QQ for chatting, while some of their children used Qzone, a social networking site very similar to MySpace and popular among rural residents and rural-to-urban migrant workers at the time (see Wallis 2015b).

18 QQ is a chat application developed early on by Tencent, and Qzone is its social networking companion site.

19 On how *suzhi* is a form of coding human value that arose at the beginning of the reform era, see Anagnost (2004).

20 The campaign started in 2017 as an effort to transform the center of Beijing into a gleaming government and financial district (Myers 2017). However, a turning point occurred when a fire broke out in November in a residential neighborhood on the outskirts of Beijing, killing 19 people, mostly labor migrants (they did not have a Beijing household registration, or *hukou*), including several children. They were living in a building with shoddy electrical wiring (the work had been performed without a permit), a fact that was then used to legitimate a 40-day "clean

up" and "rectification" campaign in which numerous so-called "low-end" people lost not only their homes but also their sources of income (Pils 2020). Demolitions of migrant enclaves have happened in Beijing in the past (see Zhang 2001); however, the speed and severity of this particular episode received widespread condemnation on social media. See https://mp.weixin.qq.com/s/CHqJZuow-Coi63iYpCic1A and https://chuangcn.org/2018/01/low-end-population/.

21 These youth, now young adults, are distinct in coming of age in the reform era. They are often viewed as selfish, spoiled, individualistic, overly concerned with material goods, and unwilling to "eat bitterness" (*chiku* 吃苦) as their elders did (Rosen 2009). As the internet and later social media diffused, they were then seen as obsessed with technology. For a detailed discussion of the challenges they face, see Kan (2013).

22 The one-child policy was implemented unevenly across China. In rural areas, families whose first child was a girl were allowed to try to have a boy after five years. This exception was partly for economic reasons (a filial son is supposed to take care of his parents in their old age), but it also exemplifies the patriarchal concern with continuing the family line through the male child. In reality, many rural families had two or more children, and the stringent enforcement of violations (through fines, forced abortions, sterilization, etc.) also depended on locale.

23 As I discuss in more detail in chapter 4, Banet-Weiser (2018) defines popular feminism as a type of lifestyle feminism that finds its corollary in popular misogyny.

24 Affect theory encompasses a wide-ranging body of work. It can be exhilarating for the possibilities it offers for fresh scholarly insights, yet maddening for the numerous and sometimes contradictory definitions of affect that are deployed. Much western scholarship traces contemporary understandings of affect to the 17th-century Dutch monist philosopher Baruch Spinoza, who, in *Ethics* (1996) defined affect as something that increases or diminishes a body's power to act (70); hence; the commonly used definition of affect as the capacity to affect and be affected, which comes by way of Gilles Deleuze (1988). Deleuze and Guattari (1987) add that affects are "intensities" as well as "becomings" (256), by which they mean that affect moves bodies that are never stable but always in process and are thus laden with potentiality. Seigworth and Gregg (2010, 6–8) delineate eight strands of affect theory.

25 Seigworth and Gregg (2010) offer an overview of these terms, which have emerged from different theoretical trajectories.

26 Seigworth and Gregg (2010, 9–10) note that "there are no ultimate or final guarantees—political, ethical, aesthetic, pedagogic, and otherwise—that capacities to affect and to be affected will yield an actualized next or new that is somehow better than 'now.'" Such seeming moments of promise can just as readily come to deliver something worse. In a similar vein, Lisa Blackman (2012, 22) states, "Affect is materialized in ways which reveal both the potential for change and hope, as well as the more insidious ways in which populations might be governed beyond normalization."

27 Berlant wrote their work to unpack how these processes have unfolded in the United States and western Europe, where constant crises have shifted notions of citizenship and the social contract, yet Berlant's broader points also relate to transformations in China, which I discuss below.

28 Petit (2015, 177) defines digital disaffect as "a kind of hypnotic, engaged disengagement with the miasmic qualities of boredom, detachment, ennui, and malaise" that accompanies our always wired lives. He is writing in the context of digital media and pedagogy, but his argument could apply to many situations involving multi-tasking and endless scrolling of social media.

29 See Audre Lorde's (2007, 147–148) description of how, as a child sitting on a bus, her Black body evoked disgust in the White woman sitting next to her, and Hemmings's (2005, 561) discussion of Lorde's narrative. These individual experiences, of positive and negative affect, should be understood as part of larger public sentiments and politics.

30 For a rebuttal to the notion of an "affective turn" and how the phrase erases earlier feminist work on emotion and embodiment, see Hemmings (2005) and Tyler (2008). See also Gorton (2007), Koivunen (2010), Greyser (2012), and Pedwell and Whitehead (2012) for very helpful overviews and analyses of feminist scholarship on affect and emotion.

31 Brian Massumi (2002) makes a clear distinction between affect and emotion, arguing that the autonomy of affect is constitutive of its possibilities. To Massumi, affect is potential or intensity, and as such it is virtual, autonomous, nonconscious (as opposed to Freud's unconscious), and exceeds language. As soon as it is fixed in language and thought (e.g., the idea of it), it becomes an emotion, which is "qualified intensity" or affect captured (28). However, my thinking aligns with feminist theorists who do not draw a firm distinction between affect and emotion and who reject this notion of "autonomy" for its erasure of agency. As Tyler (2008) argues, "(T)he danger of embracing the autonomy of affect is precisely that this claim of affect is beyond power and is thus both uncontestable and unresistable. It is important to refuse the absolute distinction between affects, feelings, and emotions not only because the purification of affects abjects an entire history of counter-hegemonic scholarship but because affect is by definition unanalyzable and thus critically and politically useless" (88).

32 For an extended discussion of the different understandings and definitions of emotion in various disciplines, see Ahmed (2014) chapter 1 and Garde-Hansen and Gorton (2013, 30–32).

33 Ahmed (2014) uses the term the "sociality of emotions" when discussing the organization of hate, but I believe it can be used to denote how a wide variety of emotions circulate and organize individual and collective bodies. Influenced by Freud and Marx, Ahmed argues that "emotions work as a form of capital: affect does not reside positively in the sign or commodity, it is produced as an effect of its circulation" (45). Moreover, "The more signs circulate, the more affective they become"

(45). See the "Afterword" for Ahmed's clarification of her focus on "objects" rather than "feelings" as what circulate.

34 In this regard, my approach is indebted to Sianne Ngai (2005). Like Ngai (2005, 27), I understand the difference between affect and emotion "as a modal difference of intensity or degree, rather than a formal difference of quality or kind," with affects viewed as "*less* formed and structured than emotions, but not lacking form or structure altogether; *less* 'sociolinguistically fixed,' but by no means code-free or meaningless; *less* 'organized in response to our interpretations of situations,' but by no means entirely devoid of organization or diagnostic powers" (emphasis in original). She adds, "What the switch from formal to modal difference enables is an analysis of the *transitions* from one pole to another: the passages whereby affects acquire the semantic density and narrative complexity of emotions, and emotions conversely denature into affects" (27). See also Yang (2014b, 2015).

35 The Chinese understanding of embodiment correlates to some extent with the feminist understanding of affect as encompassing both cognition and felt experience/bodily sensation. Jie Yang (personal correspondence).

36 According to Ji, Lee, and Guo (2010, 156), this relationality extends to thought processes and perception in that Chinese people tend to be holistic thinkers; that is, "nothing exists in isolation; things are interconnected with each other, be this directly or indirectly." I thank Jie Yang for pointing this out to me. Of course, this statement is not meant to essentialize all Chinese people as having the same thought process.

37 Zhang (2007) adds that *ganqing* is unique in that it is "always embedded in specific dynamity of social relations" (60). Roger Ames (2011, 74) states that *qing* "is both the facticity of and the feeling that pervades any particular situation. Any perceived fact/value distinction between 'circumstances' (qingkuang 情况) on the one hand, and 'feelings that are responsive to circumstances' (ganqing 感情) on the other, collapses."

38 Other relationships include *qinqing* (亲情), or "emotional attachment between family members," *fuqiqing* (夫妻情), or "affection between husband and wife," and *youqing* (友情), or friendship (Zhang 2007, 60).

39 In his ethnographic study of Chinese migrant workers in Zambia, Wu (2020) highlights the role of Chinese everyday relational ethics and what he calls "situational affects," or how emotion, which is embedded in all social interactions, guides ethical behavior. Ethnographic studies by Szablewicz (2020) on gamers and Lindtner (2020) on maker culture and innovation are also grounded in affect theory.

40 In *Love's Uncertainty*, Teresa Kuan (2015, 214, note 7) also notes that affect is similar to the Chinese word for "influence," which "invokes a breathlike vitality."

41 The number of unread messages that appeared literally overnight in my WeChat shocked and amused my colleagues during a conference in May 2019. After one of them saw the number on my screen, several crowded around to take pictures,

which inspired me to take this screenshot. I did clarify to them, however, that the large number included messages sent by members of WeChat groups I was part of, some of which had 500 members. During my fieldwork, when I asked one young feminist activist how she kept up, she replied, "It's impossible." However, she felt she had no choice but to be in numerous WeChat groups related to her job and other pursuits. For this reason, many WeChat users "mute" groups to which they belong so that they won't see the messages.

42 For a discussion of how Chinese university students in Hong Kong did "face-work" and presented themselves differently on Facebook compared to RenRen, see Tian (2017).

43 In his "Introduction" to *Ordinary Ethics*, Michael Lambek (2010) traces two main western concepts of ethics back to Plato and Aristotle. While Plato is associated with transcendence, Lambek argues that "the ordinary might be compared to Aristotle's concept of actuality (*energia*) and the unity of means and ends" (3). He further argues that Aristotle locates "ethics first in practice and action" (7).

44 As just one of numerous examples, see Part VII of Xi Jinping's Report at the 19th CPC National Congress in 2017, in which he discusses "raising intellectual and moral values" (Xinhua 2017).

45 Although "morals" or "morality" (*daode* 道德) is used more often in everyday speech than "ethics" (*lunli* 伦理), I use the term "ethics" (following scholarship on ordinary ethics, including that done in China) to speak to the bottom-up, daily lived experiences—yet, in line with this scholarship, I also often use the two terms interchangeably. In a similar vein, in *Love's Uncertainty*, Kuan (2015) writes about "moral agency and experience" to refer to "the intermediate space between the force of social norms and moral codes, on the one hand, and the capacity of actors to deliberate about their situation and to make the effort to respond accordingly, on the other" (15). I am indebted to Kuan's brilliant monograph for inspiring my thinking about the connection between affect and ethics during my fieldwork and later analysis.

46 For a critique of Lambek (2010), see Lempert (2013), who wonders how one can so "effortlessly" locate ethics. See also Lambek's (2015a) response, namely that Lempert misreads him, and Lempert's (2015, 134) reply to this response, in which he states that Lambek ignores "the communicative labor by which the ethi-cal is made intersubjectively relevant, socially recognizable, and pragmatically consequential." Similarly, Zigon (2014, 751) states that Lambek "provides us with an approach for locating ethics but not for recognizing it." In subsequent work, Lambek (2017, 139) writes, "I take ethics to be found at the conjunction of practi-cal judgment and performative action. By 'immanent' I signal that the ethical is a constitutive dimension of social life, neither transcendent of nor a detachable part of it." He adds that the ethical is based on illocutionary acts and performative ac-tion; i.e., at its most basic, do what you say you are going to do and abide by rules and commitments to which you have agreed (in marriage, religion, etc.); hence, his argument that "the immanence of ethics rests on the relationship between

practice and performative acts" (146). For an overview of these varying views of ethics, see Laidlaw (2017). I am interested in such debates, which are primarily between anthropologists of ethics who are well-versed in linguistic anthropology and language philosophy. However, determining who is "right" is a bit removed from my concern, which is not to be a theorist of ethics but rather to use ordinary ethics as a tool for analysis.

47 Xi's realization of this Dream has manifested in numerous ways, including his long-running anti-corruption drive, his tightening of ideological control, his promotion of "Chinese values," and the bolder stance China has taken in global affairs. See Gow (2017) for a detailed discussion of how the Core Socialist Values are what he calls a form of "Chinese hegemony with Confucian common sense" (108).

48 On therapeutic governance in a different context, see Pupavac (2001).

49 David Shambaugh (2021) discusses how during the Hu-Wen era, the riots that occurred in Tibet in 2008 and in Xinjiang in 2009 coincided with senior leadership changes that left conservative party members with the dominant hand. They effectively isolated the reformist Wen Jiabao and thus were able to consolidate power in the military as well as propaganda and security organs. The 2011 Arab Spring also greatly alarmed the Chinese leadership and led to more tightened online controls.

50 Neoliberalism has almost become an empty signifier at this point, but most agree that it is characterized by the confluence of the spread of advanced capitalism, new technologies of time-space compression, and more powerful international financial markets and governance entities (Harvey 2005; Brown 2015; Duggan 2003). In the West, it is synonymous with free trade, privatization, deregulation, and the weakening, if not outright discarding, of the social safety net. It also is connected to discourses about freedom, personal responsibility, and individual choice, and an ethos where consumption has become a substitute for politics (Duggan 2003). Non-state actors, including entertainment media, religious institutions, and professional fields such as psychology, also mold people's choices and define their needs in ways that align with neoliberal values (Dean 2010; Ouellette and Hay 2008; Rose 1999). "Free" neoliberal subjects are encouraged to optimize their bodily capacities, and such subjects are said to have equal access to success in neoliberal meritocratic ideology as long as they are appropriately entrepreneurial (Rose 1999). Neoliberal governance is often described as a retreat of the state when in reality the state, through its policies, creates the conditions for neoliberalism. In China, along with evolving marketization and competition, the state's hand has always been quite present in forming economic policy and controlling those SOEs that are considered strategic (oil, coal, lumber, defense, etc.) and in efforts to shape subjectivities through propaganda campaigns to promote, among other things, the former one-child policy, high levels of *suzhi*, and patriotism. Earlier scholarship debated how much forces emanating from the West, in particular "neoliberalism," could be mapped onto the transforma-

tions that China had undergone in the first few decades of reform, with scholars arguing for and against the use of this term. Andrew Kipnis (2008) cautioned against seeing China's "audit" culture as neoliberal, while David Nonini (2008) argued that China did not follow a "strong" form of neoliberalism although he did acknowledge that it might follow a "weak" form. His critique of universalist applications of neoliberalism by certain scholars was an important intervention although I think he glossed over the nuances of some of the anthropological work (primarily that of Ann Anagnost, Lisa Rofel, and Yan Hairong) that he critiqued. Other scholars, such as Gong and Yang (2017), have used neoliberalism as a way to understand China without much questioning. For a helpful discussion on how much or little neoliberalism can be applied to China, see Wielander (2018, 3–4).

51 The uber rich and middle class have benefited from these policies, while migrant laborers, many rural residents, and unemployed or underemployed college graduates lead lives defined by uncertainty and economic insecurity. In 2015, the government launched a drive to eradicate extreme poverty in rural areas, and in November 2020, Xi declared that extreme poverty had been eliminated. However, the urban poor were not the target of these measures, and it is not clear how sustainable this goal is due to the extreme measures taken in some areas to eradicate poverty (Chitwood 2020). On July 1, 2021, coinciding with the 100th anniversary of the founding of the Chinese Communist Party, Xi stated that "a moderately prosperous society in all respects" had been achieved (Xinhua 2021). During Xi's tenure, income inequality has decreased, but wealth inequality has not. As of 2021, China had the second largest number of millionaires in the world (the United States had the largest) (Credit Suisse 2021). After first mentioning "common prosperity" during an August 2021 speech, Xi later outlined a "common prosperity" policy in which he pledged to regulate high incomes to reduce inequality (Xi 2021). Common prosperity, then, seemed to be the impetus for crackdowns across a range of industries (see endnote 10 above). However, within a relatively short period of time, the phrase somewhat faded. See www.chinafile.com.

52 Such a shift had started to occur earlier and was particularly pronounced in the 12th Five-Year Plan (2011–2015), in which the word "innovation" displaced the word "reform," which had been a pillar of state policymaking since the "reform and opening" (*gaige kaifang* 改革开放) policies that Deng Xiaoping had initiated in the late 1970s (Keane 2013, 99, 103). At the Third Plenum of the 18th Party Congress in November 2013, although the economic agenda touted the role of the market in the economy, language in the agenda also insisted on the primacy of public ownership and the leading role of the state sector (See Economy 2018, ch. 4).

53 Lindtner (2020) argues that the state's desire to transform China into a nation at the cutting edge of technological innovation could only happen if China's citizens could similarly "upgrade" in the process; hence, the government discourse that innovation is for everyone. I discuss these policies in more detail in chapter 1.

54 See Hoffman (2010) for how the goals and choices of the white-collar workers she studied in the 1990s and early 2000s in Dalian were informed by professional norms and standards as well as patriotic values and ideas about strengthening the nation that were dispersed through modes of governance that could be either top-down or more diffuse.

55 For a compilation of Xi's comments on the role of women between 2013 and 2015, see www.xinhuanet.com.

56 For a popular critique of such schools, see https://mp.weixin.qq.com/s/whRNgB-49VNse2F6x09wDyg.

57 Deng Xiaoping's Southern Tour occurred in 1992, when he traveled to parts of southern China and gave speeches encouraging marketization. The presence of these speeches in state media signaled that, after the behind-the-scenes power struggle that ensued between hardliners and reformers in the wake of the Tiananmen Square massacre, the reformers had prevailed.

58 The authors note that, among older generations, this "divided self" could also result from the necessity of burying past traumas, such as experiences during the Cultural Revolution. They call these processes in which the divided self emerges the "remaking of the moral person."

59 Contrasting "a regime dominated by structures of sovereignty to one ruled by techniques of government" (101), Foucault (1991, 95) states that "whereas the end of sovereignty is internal to itself and possesses its own intrinsic instruments in the shape of its laws, the finality of government resides in things it manages and in the pursuit of the perfection and intensification of the processes which it directs; and the instruments of government, instead of being laws, now come to be a range of multiform tactics." He adds, "In contrast to sovereignty, government has as its purpose not the act of government itself, but the welfare of the population" (100). Mitchell Dean (2010), explicating Foucault, states, "*Government is any more or less calculated and rational activity, undertaken by a multiplicity of authorities and agencies, employing a variety of techniques and forms of knowledge, that seeks to shape conduct by working through the desires, aspirations, interests and beliefs of various actors, for definite but shifting ends and with a diverse set of relatively unpredictable consequences, effects and outcomes*" (18, italics in original). See also Gordon (1991) for an extensive discussion of governmentality and how this work emerged after Foucault's focus on disciplinary power. See as well Zhang (2011) for an in-depth analysis on the potential application of such ideas to governance in China.

60 All of these scholars were informed by Foucault's notion of governmentality, as well as Nikolas Rose's (1999) "governing at a distance" and his furtherance of Foucault's ideas regarding the "enterprising self." While the objects of their analyses differed, all noted a combination of a "socialist market economy," seemingly neoliberal strategies of governance, and authoritarian control. Of course, in the case of the one-child policy, which Greenhalgh and Winckler (2005) examine, women could not make autonomous choices regarding how many children to

conceive, but the way the state "conducted" how it sought compliance differed radically between urban and rural areas. My previous work (2013a, 2013b) was influenced by this body of scholarship, as I used Foucault's (1988) technologies of power and technologies of the self as a theoretical framing to unpack how young rural-to-urban migrant women used mobile communication to navigate their lives in the city. It is noteworthy that all of these works were published prior to Xi Jinping's ascendance to power. Yan Yunxiang (2009, 2011, 2013, 2017) has done extensive research on the greater individualization of Chinese society, or an emphasis on self-development, self-reliance, and competition, which has occurred as people have been able to free themselves from the collective (the rural farm, the work unit, the extended family). According to Yan, with these changes has also come greater individualism, where people have more personal choice and are freer to pursue their individual desires. However, Yan (2009, 2013) argued that unlike in the West, because this individualism is not rooted in liberal governance with its notions of liberty, freedom, and democracy, too often in China it has manifested as individual selfishness, greed, and disregard for the well-being of those outside one's familial and social (*guanxi*) circle. Thus, rather than call this the neoliberal "enterprising self," Yan (2013) termed it the "striving individual." However, as all of the authors noted above show, neoliberalism manifests differently across the globe, including in places where there is not a history of western liberal democracy. Moreover, as Wendy Brown (2015) argues, neoliberal economic policies have gutted many democratic principles such that "individual liberty" is reduced to "market freedom." This suggests, to me anyway, that liberal democracy does not need to be the foundation of the enterprising self. In slightly later work, Yan (2017) acknowledges the "desiring individual" as an aspect of Chinese personhood. More recently, Palmer and Winiger (2019) have put forward the notion of "neo-socialist governmentality" to denote Xi's tighter grip on power. In a different vein, in *Illiberal China*, Daniel Vukovich (2019) theorizes China as being "illiberal" as a contrast with (and thorn in the side of) liberalism. His analysis is grounded in intellectual/ideological debates emerging from political philosophy and international relations that are beyond the scope of this project.

61 As I discuss in more detail in chapter 2, this impetus is in the context of several state rural revitalization policies.

62 With increasing financialization of capital, Michel Feher (2018) has argued that in western post-neoliberal society, the same forces that have made shareholder profit the primary goal of institutions in turn influence everyday life. Ordinary people, especially those in a precarious financial state who must rely evermore on commercial credit, are subjected to and subjectified by a "speculative criteria of ratings and rankings" (quoted in Callison 2019). He calls this process the "economization of social life." What Silvia Lindtner (2020) notes about young tech workers in innovation hubs and maker spaces, where there has been a shift from entrepreneurialism to economization, is true of all. China's social credit system, with its

rankings and speculative value, has the potential to be an extreme version of such economization.

63 In addition to his anti-corruption drive, which initially garnered widespread popular support and also helped him consolidate power, Xi has increasingly centralized his power through creating and occupying various leadership positions. In 2013 he created and became the head of the Central Leading Group on Comprehensively Deepening Reforms and took control of the central leading groups on information technology, internet security, and foreign affairs.

64 Sometimes called "Xism," it was enshrined as Party doctrine into the Constitution during the 19th National Congress of the CCP in October 2017. The meeting is held every five years in Beijing and is attended by delegates from across China, along with Politburo members and other high-ranking officials. By having his name attached to the doctrine, Xi has ensured that no one will be able to challenge his authority (Buckley 2017; Garrick and Bennett 2018). Through this maneuver, and after already being deemed the "core" leader a year earlier, Xi put himself in league with Mao Zedong, the only other leader to have a "thought" attached to his name. For an analysis of Xi in comparison to previous People's Republic of China leaders, see Shambaugh (2021).

65 He was re-elected to a third term as General Secretary of the CCP at the 20th National Congress in October 2022 and as Chairman of the Central Military Commission and President at the National People's Congress in March 2023.

66 Many of the tenets of "Xism" look like a carryover of previous Party doctrine (e.g., the Party's supremacy, the emphasis on a people-centered approach, deepening reform) (see "His Own Words" 2017). The app, Xuexi Qiangguo (学习强国, or Study the Powerful Nation), was created by the CCP Publicity (Propaganda) Department and combines articles, quizzes, videos, and the like designed to spread Party ideology and in particular Xi's thought (Liang, Chen, and Zhao 2021).

67 Of course, across reform-era regimes, maintaining social stability has always been seen as of the utmost importance, but these changes represented a significant shift from prior years.

68 None of my research participants paid much attention to the social credit system although that does not mean it won't affect them in the future.

69 See "The Ideal Chinese Husband" (2016). Some previous leaders have also been called by affectionate terms, such as former Premier Wen Jiabao, who was known as "Grandpa Wen."

70 Yang and Tang's (2018) expanded definition of positive energy is "the capacity to induce positive emotions and/or attitudes, [and] the potential to induce constructive/conciliatory discourses and/or actions, in individuals or collectives such as the society and nation. Those positive emotions/attitudes/thoughts so induced are also simply referred to as positive energy, as is any event/discourse that is said to contain positive energy" (15). Bandurksi (2014) notes that positive energy is set in contrast to "negative energy," such as the kind found in internet memes critical of the government.

71 *Positive Energy* was also used as the title of the 2012 Chinese translation of British pop psychologist Richard Wiseman's book *Rip it Up*. The phrase went on to permeate television reality/variety shows, books on self-development and business strategy, songs, films, and government discourse. On how the party-state used it in directives to regulate internet content and in public service announcements on public transportation, see Hird (2018).

72 University students that I interviewed at the time (not included in this book) also told me they wanted their Weibo and Renren posts to be "positive."

73 See Creemers and Trilio (2022) for a discussion and translation.

74 For an insightful discussion of WeChat's infrastructural properties, see Plantin and de Seta (2019).

75 Sara Ahmed (2010) argues that the pursuit of "happy objects" is supposed to turn people's attention away from the state's responsibility for myriad problems, such as economic precarity, inequality, state-sanctioned violence, and a lack of social services.

76 Despite Mao's repudiation of "feudal" Confucian ideas, the collective was still emphasized, both in the urban work unit and the rural commune. Both Confucian and Maoist governance ultimately connected the collective sentiment to the happiness of the nation.

77 Surveys of happiness in China consistently show that unemployment and discarding of the social safety net are correlated with unhappiness (Wielander 2018). These are also related to age and education, meaning the laid-off workers of the nineties felt this most severely. Having Chinese friends whose family members suffered through these structural transformations, I can attest to how profoundly it affected them, not only economically but also morally and spiritually, especially as the family members in China had to rely on remittances from relatives working in the United States. For this reason, in China, greater economic development has not been correlated with happiness. Of course, COVID-19 lockdowns and the government's prolonged Zero-COVID policy also stirred much anger and resentment.

78 Li Zhang (2015) also notes that middle-class professionals seek out a blend of scientific (psychological) and spiritual (Buddhist, Taoist) training to "feel good," yet the highly commercialized nature of such training raises doubts as to whether it will bring long-term well-being (328). Zhang further notes the role of intellectuals and the government in these trainings (320) and that these are not just therapeutic but spiritual (321). I have several middle-aged urban friends in Beijing, whom I have known since the early nineties, who have become Buddhists over the last couple decades as a way to feel peace and contentment. As I discuss in later chapters, some of my informants also turned to religion.

79 See Yan (2011; 2020) and Li (2015) for historical overviews of the causes of the perception of a moral crisis. See also Osburg (2016) on cynicism, a loss of morality, and a "gray world," exemplified by the "*ernai*" phenomenon, or the second wives

and mistresses of Chinese businessmen with wealth and privilege, and criminal gangs who are in cahoots with state officials and the police.

80 In a different context, Lisa Hoffman (2010) recognized the legacy of socialist ethics in what she calls "patriotic professionalism," where young college graduates, who have autonomy of job choice that was denied their parents, are nonetheless supposed to serve the nation through their employment.

81 In Michel Feher's (2018) words, people must "make themselves valuable, either by advertising highly prized skills and an appealing address book, or failing that, by displaying unlimited availability and flexibility" (18).

82 See note 60 above.

83 Like elsewhere, in China social media is the prime platform for the "popularity principle" (Van Dijck 2013) and the "economies of visibility" (Banet-Weiser 2018). These imperatives, which conflate existence with visibility (Banet-Weiser 2018; Hearn 2017), are the means through which social media platforms generate value and become a site for the logics of the economization of life itself.

84 When Sina Weibo started operation, at first it was primarily used by educated, relatively young, urban internet users (including a large number of public intellectuals and celebrities, whom Sina aggressively courted to become part of its user base) (Svensson, 2014). For overviews of the development of Weibo, see Harwit (2014) and Negro (2017, chap. 6).

85 After Weibo's early "heyday," targeted crackdowns on influential users ("Big Vs") and increasing censorship diminished its ability to be used as a platform for dissent and government accountability (Svensson 2014; Han 2016). This type of usage still exists, but it is often overshadowed by celebrity news and entertainment on the platform (Benney and Xu 2017). This shift, combined with the rise of WeChat, caused some to argue that Weibo was in permanent decline (Benney and Xu 2017), yet it has nonetheless had a resurgence. As of September 2019, there were 497 million monthly active users (MAUs) and 216 million daily active users (DAUs) (Weibo Corporation 2019). These figures were both more than double those reported at the end of 2015 ("Weibo Reaches 100 million" 2016). By the first quarter of 2023, Weibo had 593 million MAUs and 255 million DAUs (Weibo 2023).

86 Compared to Weibo, WeChat was initially designed to be a much more personal form of communication, with an emphasis on sociality and entertainment (Harwit 2017; Nie, Fu, and Cheng 2013). However, this is no longer the case; adding a person to one's WeChat has now supplanted exchanging business cards, but users can overcome this less personal aspect by grouping friends and acquaintances into "circles."

87 Like Weibo, WeChat was initially used by an urban, educated group. However, unlike Weibo, WeChat was marketed to everyone, and it quickly spread. As of 2021, WeChat had 1.24 billion active users (Iqbal 2021).

88 With WeChat Pay, users can make all kinds of payments (restaurant bills, utilities, cabs, etc.). Its "mini programs" allow users to shop, book plane or train tickets or

hotels, donate to a charity, hail a ride, etc. WeChat eventually allowed companies and organizations to create public (or official) accounts to which users could subscribe. In addition to its initial outreach to all segments of users and its ease of use, Chen, Mao, and Qiu (2018) argue that it is WeChat's "super stickiness" that accounts for its popularity. To demonstrate what they mean, they begin their book with a case study of a WeChat user who took a "sabbath" from WeChat for 12 hours and found that he could barely function—he had no means of buying something at a store unless he went to the bank to withdraw cash, he could not get in touch with some people because he only had their WeChat contact information and not their mobile number, and he had no way to pay his utility bill.

89 *Wanghong* is short for *wangluo hongren* (网络红人), or literally internet red (meaning red hot) people. For a granular breakdown of such content, see Craig, Lin, and Cunningham (2020, 139). Video content and livestreamed content is viewed by roughly 95 percent and 68 percent of all Chinese internet users, respectively (CNNIC 2022).

90 For years, control over what circulates online has been achieved in numerous ways: through a system of keyword filtering and blocking, officially known as the Golden Shield Project and colloquially called the Great Firewall, and through paid as well as volunteer commentators, the latter known as the "50 Cent Army" (*wumaodang* 五毛党), who try to steer online discussions in ways that align with Party ideology (Han 2018). King, Pan, and Roberts (2017) argue that many "online commentators" are paid government employees who contribute this content (448 million posts a year) in addition to their regular jobs, not ordinary people getting paid on a piece by piece (fifty cent) basis. Official sites like the *People's Daily* (*Renmin Ribao* 人民日报) also include an online forum in which the government attempts to guide public opinion. There have also been various state-led "Anti-Vulgarity Campaigns" (Lam 2010; for a gendered critique of efforts to protest these campaigns, see Wallis 2015a). As well, the government is legitimately concerned about misinformation/rumors spread online, but also spreads its own (Roberts 2018).

91 The China Internet Network Information Center (CNNIC) releases semi-annual reports that track the numbers of internet and smartphone users, the most common activities online, rates of e-commerce, etc. In 2013, when I started the bulk of this fieldwork, China had 591 million internet users, and 165 million (nearly 28 percent) of them were rural residents (CNNIC 2013). By the summer of 2019, during my last fieldwork trip, there were 854 million internet users (with an overall penetration rate of 61.2 percent), and 225 million (26.3 percent) were rural residents (CNNIC 2019). Along with urban-rural disparity, there had been a persistent gender imbalance in internet use. In 2013, the male-female ratio of internet users was 55.6:44.4; in 2015 it was 55.1:44.9 (CNNIC July 2013, 2015). However, by 2019 the overall gender ratio had come close to reflecting the sex ratio of China's population, with 52.4 percent male and 47.6 percent female (CNNIC 2019). In terms of age, in June 2019, those aged 10 to 39 accounted for

69.2 percent; 40–49 accounted for 17.3 percent; 50–59 accounted for 6.7 percent. By June 2020, China had 940 million internet users (with an overall penetration rate of 67 percent), and 285 million (30.4 percent) were rural residents; the overall gender ratio had come closer to parity with 51 percent male and 49 percent female, and the percentage of users over 50 had increased to nearly 23 percent (CNNIC 2020).

92 In 2014, Xi created the Central Leading Group for Cybersecurity and Informatization (later changed to the Central Cyberspace Affairs Commission in 2018), which he chairs, and which oversees internet policy, security, and censorship (Roberts 2018). Just under this entity is the Cyberspace Administration of China (CAC) (also called the Office of the Central Cyberspace Affairs Commission). The CAC has implemented a range of policies designed to squelch internet freedom. For an overview and analysis of the changes, see Creemers (2017, 2021), Herold (2018), and Miao, Jiang, and Pang (2021).

93 See also the review articles by Herold and de Seta (2015) and Hu and Chen (2022).

94 For example, Papacharissi (2015, 6) has examined a particular platform, Twitter, to explore how discourses around contentious issues, such as Occupy and the Arab Spring, give rise to "affective publics," or "online and offline solidarity shaped by the public display of emotion." For earlier work on mobile communication, intimacy, and emotion, see Lasén (2004), Vincent and Fortunati (2009), Hjorth and Lim (2012), and the entire special issue of *Feminist Media Studies* 12, no. 4 (which Hjorth and Lim edited).

95 Perhaps such work is summed up best through Adi Kuntsman's (2012, 3) notion of *affective fabrics* of digital cultures, or "the lived and deeply felt everyday sociality of connections, ruptures, emotions, words, politics and sensory energies, some of which can be pinned down to words or structures; others are intense yet ephemeral." For empirical analyses focusing on students in educational contexts, see Handyside and Ringrose (2017) and Petit (2015).

96 Thus, this book picks up theoretically where my last book left off. In *Technomobility in China: Young Migrant Women and Mobile Phones*, I examined young rural-to-urban migrant women's engagement with mobile phones, drawing on a Deleuzian notion of assemblage to analyze multiple practices, meanings, discourses, and investments (financial, temporal, emotional). However, although I wrote about self-care and technologies of the self, I did not explicitly focus on the affective and ethical dimensions of such use.

97 Renren, which looked like a Facebook clone, was extremely popular among urban high school and college students in the early to mid-2010s. On its rise and fall, see Liang (2018). In 2013, while I was a visiting scholar at the Communication University of China in Beijing, all of the college students I met were using Renren. Two years later, none of them used it, preferring WeChat instead. A few years later, freshman college students were into QQ because WeChat was now for "old" people.

1. MARGINALIZED YOUNG CREATIVES, PERSONAL AESTHETICS, AND THE QUEST FOR MEANING

1 All names in this chapter and throughout the book are pseudonyms unless otherwise indicated.

2 Fieldwork in Beijing, September 2015.

3 In the past, China's rigid household registration system severely constrained population movement inside (and outside) the country. In recent years, there have been a number of national-level reforms that have removed obstacles for rural residents in smaller cities to convert their *hukou* from rural to urban. In larger cities, such policies still tend to favor the educated, "talented," and wealthy. In mega-cities like Beijing, the policies remain the most restrictive. See Wallis (2013) chapter 1 for an extensive overview of China's *hukou* policies up to the early 2010s. For a discussion of more recent *hukou* reform, see Alpermann (2020).

4 A *pingfang* home is a flat, one-story house, often in a *hutong*. Although *hutong* living is often romanticized by westerners, most *hutong* homes are rundown and don't have adequate heating or air conditioning. Many also don't have bathrooms, so tenants must use public restrooms. Having lived in a *hutong*, I can attest that the experience was not very pleasurable. *Hutong* homes that have been fully renovated are far too expensive for ordinary Chinese people (and scholars on a tight budget) to afford.

5 Such efforts were spurred on initially by scholars like Richard Florida (2002) and his work on the "creative class" and "creative cities," and by UK government initiatives promoting the "creative and cultural industries (CCI)" (Hesmondhalgh and Baker 2011). Florida's overly positive assessment of the role of creatives in transforming cities received several critiques (see Peck 2007). In later work, Florida (2017) acknowledged the problems with his creative class thesis and the inequality that had increased in major cities where the focus on creativity led to gentrification and displacement.

6 The findings in this chapter are based on ethnographic fieldwork conducted in Beijing for varying lengths of time in 2013, 2014, 2015, 2017, and 2019 as well as digital ethnography. Although I met numerous young creatives during those years, this chapter draws primarily from interviews, several with multiple follow up, and casual conversations with 15 marginalized young creatives: eight men and seven women living in Beijing. They had all been born in rural towns or villages, and thus had a rural household registration (*hukou*). Some had completed a year or two at a provincial university before dropping out, and several had two- or three-year vocational degrees in various fields, meaning they had more education than most labor migrants. When I first met them, they ranged in age from 22 to 27 years (except for one man, Hu Fang, who was in his mid 30s). During the research, I established long-term relationships with 65 key informants, solidified through casual conversations over meals or coffee, attending various artistic/creative events (art exhibits, gallery openings, musical performances, film screenings,

etc.) together, and messaging via social media. Participants engaged in differing amounts of posting on social media. Analyses of a subset of participants' WeChat posts are the result of in-person sharing, chats via social media, and intermittent digital ethnography during parts of the latter fieldwork and into mid 2020.

7 Like those of their western counterparts, these studies (with the exception of L. Zhang [2023]) tend to focus on creative workers who are college-educated, or in the process of obtaining a degree, and employed in particular industries (art, tech startups, television).

8 In these longer-term efforts, in state and popular discourse, urban, educated Chinese are assumed to have the capacity for ethical self-cultivation as they embrace middle-class subjectivities, while rural residents and migrant workers are viewed as always starting from a position of lack (Anagnost 1997; Tomba 2014; Wallis 2013; Yi 2019; Zhang 2020).

9 My approach is similar to that of Chow Yiu Fai (2019), who, while acknowledging how affect and emotion have been key to the exploitation of creative workers (through, for example, discourses of freedom and doing what you love, which tap into feeling and emotion while veiling the instability and exploitation embedded in what is often precarious work), focuses on the meaning and rewards such work can also bring, and not in ways that are only synonymous with "cruel optimism." In an earlier article, Chow (2017) addressed ethical issues that arose when a participant in his study was asked to do something unethical in the course of their job.

10 These data are from interviews conducted in 2013 and 2015 with 9 urban, college-educated young creatives (four men and five women, aged 22–28) employed in corporate settings in creative industries, including advertising, marketing, graphic design, and state media.

11 Exhibits at the World Expo also displayed how many people had been lifted out of poverty, China's rate of urbanization, and the nation's technological advancements.

12 Of course, fakes can also be deadly, such as tainted milk powder, lead in toys, and melamine in dog food. See, for example, Spencer (2008).

13 "Innovative nation" appeared in China's 11th Five-Year Plan (2006–2010), and the word "innovation" became more pronounced in the 12th Five-Year Plan (2011–2015), replacing the word "reform," which had been foundational in state policymaking since the "reform and opening" that started in the late 1970s (Keane 2013). The 13th Five-Year Plan (2016–2020) also made innovation key to China's development strategy and added "mass entrepreneurship." The year 2015 saw the arrival of the Internet Plus plan and Made in China 2025 (See the introduction, endnote 8). This focus on innovation is also intended to improve China's soft power abroad (Keane 2013; Wallis and Balsamo 2016; Yang 2016). TikTok, the English-language version of Douyin, seems to be the biggest success in this regard thus far.

14 Michael Keane (2013) explains that the term cultural industries is more aligned with national state policy and emphasizes the role that these industries and institutions play in propagating the "great civilization" discourse. Cultural industries also include "tourism, publishing, advertising, design, arts and crafts, broadcasting and digital media" (27). Creative industries is a term imported from the West (particularly the UK and US) and highlights creativity, originality, and iconoclasm. Keane notes that these values do not naturally coalesce with an authoritarian state that wants "harmony" of ideas. Often these two terms are collapsed as "cultural creative industries" (*wenhua chuangyi chanye* 文化创意产业).

15 In Beijing, art districts include 798, CaoChangdi, and Songzhuang. On maker spaces and design houses in Shanghai and Shenzhen, see Lindtner (2020).

16 Lindtner (2020) explains how a transnational maker movement, born out of a US libertarian belief that by enabling anyone to have access to technology production, equality and freedom would also follow, emerged in China. These ideals were first embraced by a small network of makers and hackers in cities such as Shanghai and Shenzhen. However, after Li Keqiang visited a maker space in Shenzhen in 2015, the government appropriated ideals attached to making—empowerment, agency, DIY, and so on—in service of its own desire to transform China into a nation at the cutting-edge of technological innovation. This transformation could only happen if China's citizens could similarly "upgrade" in the process; hence, the government discourse that making was for everyone. Lindtner argues that, as a result, practices around making furthered processes of the economization of life, as making was appropriated as yet one more domain for human life to become human capital for finance capitalism. Lindtner (2020) also shows how discourses associated with the maker movement simultaneously reify China as a copycat nation and fetishize China, particularly Shenzhen, as the next frontier of technological innovation. Both Lindtner and Fan Yang (2016), in her analysis of global intellectual property rights in *Faked in China*, reveal how, in trying not to copy but to innovate, China still was seen by some as copying the West.

17 China's 14th Five-Year Plan (2021–2025) includes an emphasis on innovation and cultural soft power as two of several key priorities. See the DigiChina translation here: https://digichina.stanford.edu.

18 For an insightful overview of the origin of and a critique of precarity from different intellectual and activist traditions in the context of western capitalist countries, see Gill and Pratt (2008).

19 As the industrial base in many western countries was unraveling, China was undergoing a new industrialization in the 1980s, powered by thousands of migrant workers leaving villages, where they had never been supported by the state, to do exploitative labor in factories. Many rural residents have experienced economic insecurity and engaged in monotonous labor. While previous generations knew only farming, and then later migrant labor, many younger rural residents have more opportunities than their parents and grandparents.

20 In his study of single women from diverse backgrounds (in terms of age and experience) working in a variety of creative fields in Shanghai, Chow (2019) discusses the challenges they face, including that some are underpaid, lack job security, and must contend with sexist discourses to pursue their passions. He rejects the term "precariat," however, in order to highlight their agency rather than their victimhood. Drawing from Hesdmondhalgh and Baker's (2011) study of creative workers in the UK, he frames his analysis in terms of how these women construct "good" work, which offers autonomy, meaning, and opportunities for personal and professional growth. Like Chow, I also wish to highlight the agency of the young creatives in this chapter, but my focus is on how they framed their work in terms of personal aesthetics and ethics.

21 This form of self-cultivation is more regulated than what I am discussing in this chapter.

22 The preeminent example is found in the work of Chinese modern literary giant Lu Xun. On ethics in contemporary Chinese literature, see Visser (2010). On the meaning of culture during this time period, see O'Connor and Gu (2020).

23 See McDougall (2020).

24 Of course, Chinese artists in the reform era have always had to walk a fine line between "renegade" expression and not rocking the boat. When I lived in Beijing in the early 90s, just after the Tiananmen massacre, events with even a minimal oppositional stance, such as a Cui Jian concert, were usually publicized only by word of mouth, and they were just as likely to be canceled as to take place. The same is true now, but the prominence of social media and the government's surveillance state make it even harder for them to remain under the radar.

25 For a breakdown of the speech, see www.chinafile.com. In the years since, there have been numerous crackdowns on and censorship of an array of pop culture forms, including rap music, fandom, and so-called "sissy" and "abnormal aesthetics" of male pop idols (see National Radio and Television Administration 2021).

26 The young rural-to-urban migrant women featured in my first book almost unanimously repeated a similar discourse. See Wallis (2013) chapter 2.

27 Fieldwork in Beijing, October 2015.

28 For a popular take on *Yuanmingyuan* at the time, see Tefft (1993).

29 Lin (2023) uses the term "bilateral creatives" to denote how those employed in SOCEs need to balance the ideological and commercial while striving for self-realization and autonomy. Hu Fang's departure from the SOCE enabled him more creative freedom.

30 Fieldwork in Beijing, April 2013.

31 In previous research, I observed how, among young rural-to-urban migrant women, access to the internet, and later social media, opened up spaces for new modes of subjectivity and ways of imagining possibilities for one's life. However, the ambitions expressed by these women were humbler and focused on the practical goal of "studying" in the city, which could mean anything from modifying

their appearance to look less rural to improving their literacy, computer, or job skills so that they could obtain better employment. While doing fieldwork for this book, the labor migrants whom I met as I went about my daily life, who were employed in conventional fields such as service industries, differed as well from earlier "second-generation" migrant workers. In casual conversations, these more recent labor migrants seemed to have more of a sense of their own ambitions and were much less likely to say they had come to Beijing to "see the world" or "gain some skills," as the young rural-to-urban migrant women I knew in Beijing several years earlier had. See Wallis (2013).

32 Fieldwork in Beijing, April 2013.

33 On boredom as motivation for creativity in China, see de Kloet, Chow, and Scheen (2019), especially the "Introduction."

34 Fieldwork in Beijing, March 2013.

35 Fieldwork in Beijing, April 2013.

36 Although the "ant tribe" phenomenon still exists and is a concern for the government, this term is not used as much now.

37 Fieldwork in Beijing, September 2015.

38 As Chumley (2016) notes, when the government latched on to the creative economy, art institutions created new departments, and new institutions came into being, meaning that more students than ever began studying art, particularly graphic design. The university entrance exam scores needed to get into art departments are also lower than for other fields. Chumley states that, for this reason, none of her informants studied art because it was their passion. Rather, based on their subpar academic performance in junior middle school, teachers and administrators urged them to pursue an art track; and parents, particularly those from rural areas and/or lower socioeconomic strata, supported such a decision, seeing it as a way for their children to gain entrance to university. For these reasons, during fieldwork in Beijing and rural Shandong, I met an abundance of students majoring in graphic design as well as recent graduates with design degrees.

39 Fung (2016), in his comparative analysis of creative labor in East Asia, finds in China's creative clusters those who have talent but are constrained by a corporate culture that is "pragmatic rather than imaginative or spiritual" and who eschew rocking the boat as long as they can make a decent salary and lead a "settled" life (211). He contrasts these workers with "progressive artists" in the US and South Korea. He also states that "despite the 'creative' nature of the industries, the programmers, artists, and marketers in these game companies resemble industrial workers in their tastes, aesthetics, and lifestyles" and their offices "resemble factories" (112). Thus, many scholars have argued that a large portion of what is truly creative in China is found online, where clever memes, homemade videos, internet novels, and the like are produced and enjoyed by millions (Herold and Marolt 2011; Wallis 2011; Yang 2009; Lin and de Kloet 2019).

40 In *The Use of Pleasure*, Foucault (1990, 26–27) distinguishes between four types of conduct in the formation of an ethical subject: "*the determination of ethical*

substance," or how one constitutes oneself as "the prime material of his moral conduct" (26); "*the mode of subjection*," or how one relates to and puts rules into practice (27); "*ethical work*," or how one attempts "to transform oneself into the ethical subject of one's behavior"; and "*telos*," or the desired end result (27–28) (emphasis in original). In a subsequent interview (Foucault 1997), the second type is translated as "subjectivation," and Foucault calls the third type "self-forming activity" (265). All of these involve technologies of the self, which Foucault (1988, 18) defines as those techniques "which permit individuals to effect by their own means or with the help of others a certain number of operations on their own bodies and souls, thoughts, conduct, and way of being, so as to transform themselves in order to attain a certain state of happiness, purity, wisdom, perfection, or immortality." These actions entail some measure of freedom, in contrast with technologies of power or domination. Although Foucault was drawing upon ancient Greek texts, his ideas have been taken up and integrated with Chinese notions of self-making and the construction of the ethical self (Kuan 2015; Wu 2020; Yi 2019; Zhang 2011). In this chapter, my focus is most often on what Foucault calls the ethical work/self-forming activity and the *telos*, and how young creatives pursued these within the constraints of the neo/non-liberal milieu.

41 Lei Feng was a soldier who died in 1962 and shortly thereafter was upheld by Chairman Mao as a role model for the entire nation due to his passion for serving the people. He has been periodically rejuvenated in government propaganda campaigns since then but has also been the source of derision in unofficial popular culture. For more on this history, see Jeffreys and Su (2016).

42 Fieldwork in Beijing, June 2013.

43 The people they followed on Weibo are too numerous to list, but well-known names included Lee Kai-Fu, the former president of Google China, and Wang Xiaoni, a famous poet.

44 Fieldwork in Beijing, October 2017.

45 Fieldwork in Beijing, May 2013.

46 Fieldwork in Beijing, October 2017.

47 Some tattoos I have seen beg the question as to whether the tattoo artist was deliberately hoodwinking their customer. A friend relayed to me perhaps the best example: a westerner she saw in Tianjin with the Chinese characters 文盲 (*wenmang*), meaning "illiterate," tattooed on his arm.

48 Fieldwork in Beijing, October 2017. The *Sida mingzhu*, from the Ming and Qing dynasties, are also called the four great masterpieces: *Water Margin*, *Romance of the Three Kingdoms*, *Journey to the West*, and *Dream of the Red Chamber*. These ideas about copying cut across education levels, social class, and rural/urban origin. For example, one middle-class informant, William, was an urban resident with a master's degree in international business from a British university. He had had a secure job at a well-known finance magazine, yet he eventually quit because it was "boring." His cultural and educational capital set him apart from the marginalized young creatives who are the focus of this chapter. With his

savings, he had bought a few apartments in Beijing and rented them out through Airbnb, mostly to foreigners (before this practice was banned). He also curated a Lofter (akin to an Instagram account, but with a primarily design focus), where he regularly posted photos of interior designs he found appealing. His account quickly gained followers and became a recommended site. As we scrolled through his Lofter, he said many of the photos were from western social media platforms such as Tumblr. He thought of this as cultivating a cosmopolitan image and said it was sharing not copying.

49 Researching the aforementioned Dafen Village, Willie Wong (2013) unpacks the false dichotomy between the supposedly "alienated labor" of the Dafen "copycat" painters, few of whom actually work in the so-called art factory assembly lines mentioned earlier, and the "creativity" and "individuality" of high-profile western conceptual artists. Many of the latter employ the same artistic practices as their counterparts in Dafen, including copying and contracting out their labor to apprentices. She also argues that the Dafen artists, most of them migrant workers, are quite unlike the image that has been depicted of them: exploited, alienated, unable to create, and victims of either a "global capitalist system or totalitarian communist state" (15). Focusing on art education in China, Lily Chumley (2016) reveals a difficulty faced by university art students. Their art prep schools provided a grueling education in technical skills and mimicry of state-approved forms, yet they were at a loss when suddenly expected to develop their own unique style and artistic persona. Both Wong and Chumley provide insights into the complicated and contradictory processes involved in creative production and the construction of the creative self.

50 Of course, poststructuralist, anti-humanist critiques upset such notions long ago, yet the durability of these ideas remains (and has formed and sustained the basis of US copyright law; see Pang 2012). It is also telling that it was not until the internet, and in particular "Web 2.0," emerged and enabled the widespread circulation, sharing, and reconfiguring of cultural products, that many of these critiques became more mainstream.

51 Sundararajan (2015) adds that a "harmony model" also means that "creative action requires going both with and against the flow" (144). In a contemporary context, these ideas are perhaps best represented by the *shanzhai* (山寨), or copycat, phenomenon that emerged in the first decade of the 2000s, originating around mobile phone production in southern China, where the confluence of flexible modes of production, translocal and cross-border cultural flows, and a dynamic working-class ICT culture led to the development of phones that were simultaneously knock-offs and original (phones shaped like race cars or that had shavers attached, and phones that contained a dual sim card) (Wallis and Qiu 2012). On *shanzhai* culture more generally, see de Kloet, Chow, and Scheen (2019). My informants rarely mentioned *shanzhai* products or practices, however. They reserved that connotation for low-quality products and/or the practices of the less educated migrant workers from whom they distinguished themselves.

52 Fieldwork in Beijing, September 2015.

53 Fieldwork in Beijing, August 2015.

54 She did post pictures of her cat, however, in contrast to Bo's stringent judgment.

55 Fieldwork in Beijing, October 2017.

56 Fieldwork in Beijing, September 2015.

57 Fieldwork in Beijing, September 2015.

58 As discussed in the book's introduction, WeChat enables users to separate contacts into different groups (family, friends, classmates etc.). Only a few of the young creatives said they did that.

59 Gu (2003) acknowledges that this concept can be compared to "postmodern conceptions of unlimited semiosis and 'openness'" (491), yet anyone who has watched early films directed by some of China's Fifth and Sixth generation film auteurs, such as Zhang Yimou or Jia Zhangke, will note the scarcity of dialogue compared to typical Hollywood dramatic films. Gu also argues that prior to the postmodern turn, such suggestiveness is what distinguished Chinese from western aesthetics.

60 Although, on the surface, this practice of providing minimal context or comment seems similar to the practices of the domestic workers I discuss in chapter 3, the underlying motivations were quite different. The domestic workers almost unanimously said they did not include many comments because they didn't know what to say or they were not skilled at writing Chinese characters.

61 He invited me to join one of these groups, but I am not at liberty to discuss what was talked about in that group.

62 Fieldwork in Beijing, September 2017.

63 Almost every Chinese person I know will say they are not interested in politics. The word is too loaded and risky.

64 See https://mp.weixin.qq.com/s/OICpqPXDRF5RyGyjIJQ4-w and https://mp.weixin.qq.com/s/QJtvGCwXl7JO6YZmhQSfyQ.

65 For a discussion of a film about *chengguan* (not Wilson's film), see Fu (2021).

66 The 996 work schedule became a controversy especially in the spring of 2019 after Jack Ma, Alibaba's founder, and Liu Qiangdong, the founder of JD.com, both (although separately) endorsed it. See https://edition.cnn.com. Ma's remarks possibly were one factor of several in his eventual censure (see the introduction).

67 Fieldwork in Beijing, July 2013.

68 Fieldwork in Beijing, October 2015.

69 In 2006, when I first visited 798 in Beijing, although it was regulated by the government, it often hosted punk shows and experimental art, and there were only a few bars and restaurants. Over time, it became so commercial and touristic that I and everyone I knew stopped going there.

70 In contrast, some of the domestic workers and rural entrepreneurs in this book did not view marketing to friends as necessarily a problem. It was perceived as sharing and helping each other rather than as instrumental commerce, as long as it wasn't excessive.

71 Fieldwork in Beijing, October 2017.

72 Although livestreaming has become a major means for both urban and rural people of all kinds to sell things or capitalize on a particular skill, none of the young creatives that I met livestreamed. On livestreaming, see Larson (2017).

73 For more on Momo, see Chan (2020).

74 The man playing the erhu was getting set up at the intersection, and Wilson was using a small handheld video camera to film him from different directions. Wilson told me he had been filming the man for several days, with his permission. He eventually edited the footage he took into a 45-minute documentary on buskers and night vendors, who face extreme harassment from the *chengguan*, or urban management officers, and submitted it to a film festival. He subsequently made a full-length documentary of a musician that was accepted to a film festival outside of China.

75 Fieldwork in Beijing, September 2015.

76 Fieldwork in Beijing, October 2017. When they first started out, Yuki and her boyfriend worked out of their home. Once, when she was working alone with a male customer, he propositioned her. When this occurred, she was very offended but also so scared that she ran outside and stayed there until the man left. Afterwards, she found a job in a tiny tattoo shop in central Beijing. Had her father known about this incident, it would have confirmed his worst fears.

77 Fieldwork in Beijing, August 2019. In this way, Leo anticipated the "lying flat" (*tangping* 躺平) phenomenon that exploded in 2021 as a reaction to and rejection of a culture of overwork (the 996 schedule mentioned earlier), and also as an expression of the general malaise and anxiety caused by COVID-19 and the Zero-COVID policy. See Ji (2021).

78 See my discussion in the introduction on Yan Yunxiang's (2010, 2013, 2017) arguments regarding individualization and individualism in China.

79 Fieldwork in Beijing, September 2017.

80 Ironically, the gallery where Xiao Sun worked had prints of Andy Warhol's Marilyn Monroe silkscreens, most likely produced in Dafen Village.

81 The two are not necessarily mutually exclusive. A very successful commercial photographer I have known since the early 1990s studies with a Buddhist master.

2. CHALLENGING TECHNOSOLUTIONISM

1 Rural development has been a focus of the government for decades. For a detailed account of early rural reform, see Unger (2002). On more recent urbanization efforts and their degree of success, see Gao and Su (2019) and Gong, Wei, and Gu (2022). The parents of my host in the village were not happy with the offer the government had made to them to give up their home and land and move to a high-rise apartment building. They felt they would be losing their financial security.

2 Micro-enterprises have nine or fewer employees (Kushnir, Mirmulstein, and Ramalho 2010). In rural China, these employees are often family members.

3 In his address at the 19th National CPC Congress in October 2017, Xi Jinping called for further rural revitalization efforts in order to eliminate rural poverty by 2020. My host attributed the physical changes to the "five revitalizations" (of rural industries, talents, culture, ecology, and political organizations) (see B. Li 2018).

4 All names of villages, participants, and the skincare company discussed later in the chapter are pseudonyms. Suan Village served as the key village where research took place during the fieldwork, and where key informants introduced me to fellow villagers and residents of surrounding villages.

5 Fieldwork in Suan Village, August 2019. Older people are more likely to have more data traffic on WeChat than younger people (most likely because younger people are using a bigger variety of apps). See www.199it.com.

6 Informatization is access to technology infrastructure, applications, and services (Qiang et al. 2009). As discussed in the book's introduction, in its current incarnation in China, it is connected to the digitalization of the economy, industry, and governance (see Creemers and Triolio 2022).

7 This chapter incorporates and builds on my previous research in rural Shandong, in which I examined how the use of communication technology for micro-entrepreneurship became the site for the reproduction and/or reconfiguration of gendered power relations (Wallis 2015). In rural China, a family business often means that several nuclear and extended family members are involved in different aspects of the enterprise. In focusing on couples, I am using the term "family business" more broadly.

8 The six research sites in Shandong reflected this diversity, as two were relatively close to Qingdao, three were in central Shandong, and one was in a more mountainous region in the south. The first, Yu Village, was about an hour drive from the center of Qingdao, and its economy was based on fishing and seafood processing. Even during my first visit, it had newer infrastructure (e.g., paved roads, restaurants, a large factory nearby) and an overall higher standard of living, as did the second site, Zhuang Village, which was two hours north of Qingdao and had its own textile factory that employed mostly local women. Villagers also engaged in agriculture. The third site, Suan Village, was located about 100 kilometers from Jinan in central Shandong. The local economy was based on agriculture, particularly the cultivation of garlic, onions, and ginger. In 2011 and 2013, there were a few small grocery stores and other shops, and the lanes of the village were paved. Many local residents had recently built new houses or remodeled old homes, but these had pit toilets and the only running water was a spigot in the outside courtyard. There were no restaurants or entertainment venues. In 2019, even though the government had invested in a beautification program, a lot of homes had been abandoned after residents moved to apartments in the nearby town or city. The fourth and fifth villages—Shi and Bai—were adjacent to Suan Village and nearly indistinguishable from it in terms of the agricultural economy. However, Bai Village was slightly more developed than Suan and Shi. During fieldwork in 2013, the main road was being repaved and the middle of the

village had a new shopping area with a two-story department store. In 2019, there was more development in the form of two-story buildings and paved roads. The sixth site, Zhu Village, was about an hour's drive south of Suan Village in a mountainous region. It consisted almost entirely of homes and small family enterprises focused on raising pigs or garlic. The roads were not paved, there was no village elementary school, and there were no businesses devoted to leisure activities. Like Suan Village, even new homes did not have plumbing aside from a courtyard spigot.

9 My trips to the Shandong countryside were facilitated by key informants who were raised in the villages but were working in urban areas. They accompanied me and, in some cases, two other researchers to their villages and adjacent villages. All my trips involved a combination of participant observation, interviews, most of which lasted from 30 minutes to one hour and some with follow-up discussions, and casual conversations. In 2011, fieldwork was conducted in five villages (Yu, Zhuang, Suan, Bai, and Zhu) in different regions of Shandong. Data were gathered primarily through semi-structured interviews with 30 informants (12 men and 18 women) who ranged in age from 18 to 50, and through participant observation at a small number of homes and worksites. All but one of the men had migration experience, and nearly all the women either had been employed at some point outside their village (returning home after work) or had labor migration experience. In 2013, I conducted follow-up research in three of the original villages (Suan, Bai, and Zhu) and Shi Village. During this trip, interviews were conducted with 14 new participants, and I was able to follow up with five people from the previous fieldwork. The 14 new participants ranged in age from 18 to 48, with nine men and six women. Of these, there were five married couples. Only three men and one woman had migration experience. Two young women were college students, and one middle-aged woman had never left the village to work. I include only a small subset of these data to contextualize the later fieldwork. In 2019, I visited Suan, Shi, and Bai villages. Many of my previous informants had left the villages. In addition to Ms. Fang and a married couple I had interviewed on both previous trips, I interviewed seven new informants (two men and five women, aged 37 to 60). Of these, one of the men and two of the women had migration experience. After the 2013 trip, I received occasional updates from a few participants via social media (QQ and WeChat), but these faded over time. After the 2019 trip, my host provided updates, which I discuss in the book's conclusion. The analysis in this chapter is based on interview data and fieldwork notes.

10 Yang (1994, 67) states that *renqing* is both "both ethical and emotional" and is what makes humans human. She also notes that demonstrating *renqing* makes one morally worthy or virtuous. There is a large body of research on *guanxi*, *renqing*, and other indigenous Chinese concepts in business and management studies (see, for example, Yen, Barnes, and Wang 2011). However, because such scholarship focuses on formal business organizations, most often in urban areas, it is beyond the scope of this chapter.

11 My use of "appropriate" means "suitable," and thus differs from "technology appropriation," in which users shape and adapt a new technology.

12 During the Mao years, the state tried to shift people's allegiance from family and kinship to the collective. Oxfeld (2010) notes that eventually the government had to allow for some allegiance to family in rural areas.

13 Of course, this blending of work is not unique to rural China; rather, it has been the norm in many agricultural societies.

14 One of the first changes in rural China when the reform and opening policy began in the late seventies was the dismantling of the agricultural collectives. The "household responsibility system" enabled peasant families to diversify their economic activities as long as they met their production quota for the state (see Croll 1987). I distinctly recall that, in the winter of 1990, while traveling with a Chinese friend by train from Hangzhou to her small hometown in Zhejiang province, she occasionally gestured out the window to large, newly built houses that dotted the countryside. She explained that many farmers had gotten "rich" as a result of the recent market reforms and that they were quite happy. This situation did not last, however.

15 In the early eighties, the government enacted a series of reforms that dramatically transformed the countryside in terms of economic activities, household autonomy, and mobility. For a brief overview, see Wallis (2013, 35–40).

16 Bray (1997), examining imperial China, and Bossen (2002) and Song (2015), in their discussions of both the Mao and reform eras, document rural women's engagement in domestic sidelines. Bray discusses weaving, Bossen mentions embroidery and shoemaking, and Song points to needlework as a "side job." The difference between the Mao and reform eras is that, during the former, women's sidelines were more likely to be recognized as an economic contribution. See also Jacka (1997).

17 In the eighties and nineties, the government emphasized town and village enterprises (TVEs) as a means of soaking up surplus rural labor. The young women who found employment in TVEs no longer had time for domestic sidelines. However, in areas where men still had the responsibility for farming, their labor was more valued than women's (Croll 1995; Song 2015).

18 Somewhat ironically, when villagers build big houses, throw lavish wedding feasts, and "waste" money on funerals to honor ancestors, these are sometimes not seen as villagers following moral codes related to face, *guanxi*, and ritual, but rather more evidence of the backwardness and ignorance of rural residents.

19 Oxfeld states that if someone says an individual has "no conscience," it means they are not a moral person. She adds that *liangxin* "embodies ideas about individual moral responsibility and the importance of memory, for to accuse someone of lacking conscience is to say that she has forgotten her obligations" (46).

20 Steinmüller (2013, 23) borrows Michael Herzfeld's notion of "cultural intimacy," or "'the recognition of those aspects of a cultural identity that are considered a source of external embarrassment but that nevertheless provide insiders with

their assurance of common sociality (2005, 3),'" to get at the difference between how villagers see themselves and how outsiders do, as exemplified in official representations. The way villagers negotiate these perceptions results in what Steinmüller calls "communities of complicity," or "communities of those 'in the know,' those who share an experiential horizon and an intimate knowledge" (224). Words such as "embarrassment" and "complicity" evoke negative connotations, although of course there are positive outcomes as well.

21 McDonald (2016) argues that moral accumulation is gained through demonstrating resourcefulness, persistence, and hard work. He offers a rich account of his participants' lives, but aside from connecting "leveling up" in an app to entrepreneurship (because it necessitates qualities such as diligence and resourcefulness), he does not discuss entrepreneurship, and gender is not part of his analysis.

22 China's eminent sociologist, Fei Xiaotong (1992, 62–63), used the term "differential mode of association" (*chaxugeju* 差序格局) to denote a pattern of social relations in which, like the circles that radiate outward when a stone is thrown into a body of water, each person "stands at the center of the circles produced by his or her own social influence." In this conceptualization of social relationships, each subsequent circle denotes decreasing levels of closeness, and thus significance, to the person at the center. These social circles are "highly elastic" (64).

23 Hwang (1987, 954) states that the three components of *renqing* are proper emotional responses, a resource that "can be used as a medium of social exchange," and "a set of social norms by which one has to abide in order to get along well with other people in Chinese society." He emphasizes the role of *renqing* in what he calls "mixed ties," meaning ties that are not solely expressive, like those between close friends and family members, nor solely instrumental, such as between a salesperson and a client. These mixed ties are particularistic and "occur chiefly among relatives, neighbors, classmates, colleagues, teachers and students, people sharing a natal area, and so forth" (952). Hwang views *renqing* as part of a power game based on receiving and allocating resources. In John Osburg's (2016) study of the "gray world" of relationships between mistresses or second wives and wealthy businessmen, and between gangsters and state officials, *renqing*, as embedded in relationships based strictly on personal interest devoid of any greater moral or ethical foundation, takes on a much more sinister meaning than in my usage of the word in this chapter, including later when I discuss Ms. Chen and "gray commerce."

24 I use these terms not to fix the villagers in an ossified culture that is not in tune with "modern" notions of relationships but rather to recognize the nuances and norms of their relational and business interactions.

25 Fieldwork in Zhu Village, July 2011. It is very common for village youth to go to boarding schools in nearby towns for junior and senior middle school.

26 On a return visit in 2013, the Lius did have internet, and the computer, no longer an object of special pride and novelty, had been moved into one of the bedrooms.

27 Numerous policy documents were then filled with solutions for alleviating the *sannong* ("three rural") crisis, the three elements being *nongcun* (countryside), *nongye* (agriculture), and *nongmin* (peasant farmers). For a list of the "No. 1 Documents" released between 2004 and 2010, see Chun (2011, 85n2). As Hu (2016) points out, gender was not a focus in the early discussions of *sannong*.

28 The phrase "New Socialist Countryside" was put forward in conjunction with China's 11th Five-Year Plan (2006–2010) and 2006 "No. 1 Document." Aside from more investment in infrastructure and education, a centuries-old agricultural tax was eliminated. While living in Beijing at the time, on brief visits to rural villages I noticed that chalkboards with handwritten explanations of the meaning of the New Socialist Countryside were ubiquitous. Subsequent Five-Year Plans have continued to elaborate on rural development.

29 For an overview of informatization policies in the 1990s and 2000s, see Harwit (2008), Hong (2017), and Wallis (2013, 53–61). By the early 2010s, landline telephony had diffused to nearly all of rural China. In 2012, China's Ministry of Industry and Information Technology (MIIT), in coordination with seven other ministries, launched a plan to provide broadband internet to villages, including in rural schools (MIIT 2013). In 2011, when I first went to the villages, according to the China Internet Network Information Center (CNNIC), China had 485 million internet users (roughly 55 percent male and 45 percent female), and 131 million were rural residents, accounting for 27 percent of the total (CNNIC 2011). Prior to and throughout the New Socialist Countryside period, numerous projects were implemented to promote ICT access, enhance entrepreneurship, support e-governance (State Council 2006; Harwit 2008; Hong 2017; Looney 2015), and develop agriculture through the use of ICTs (Xia 2010; Zhao 2008). A 2016 CNNIC report on rural China noted that by the end of 2015, there were 195 million internet users in the countryside, and 170 million of them (87 percent) accessed the internet via a mobile phone. Although more and more rural people were integrating the internet into their daily lives and into business, 68 percent of the rural populace did not use the internet (CNNIC 2016). The trend of an increasing number of rural internet users has continued. See the introduction, endnote 91 for further statistics from later CNNIC reports. The 2020 CNICC report included a special section (pp. 15–18) on technology and poverty alleviation.

30 For example, through the 2009 White Goods policy (to counter the effects of the global economic downturn), villagers (through subsidies) were encouraged to purchase consumer goods, such as mobile phones and computers as well as washing machines and refrigerators. In 2013, when WeChat had not yet widely diffused outside of cities, China Telecom promoted smartphones with WeChat in rural areas (see https://tech.qq.com/a/20130722/011625.htm). Despite such promotions, many rural residents in the 2000s and early 2010s received computers and cell phones, either new or hand-me-down, from relatives who had migrated to work in towns and cities. These were primarily used by children for school and

entertainment, and by adults for social, not economic, reasons (Liu and Ye 2010; Oreglia and Kaye 2012).

31 Fieldwork in Yu Village, July 2011.

32 For years, internet cafés were portrayed in state and popular discourse as places where teenaged boys and young men incessantly played video games, causing addiction, physical health problems, and in some cases, even death. See Szablewicz (2020, chap. 3).

33 It is possible that she was actually the one returning a favor.

34 Lin Ying relayed to me how she and Hu Yang had met in a QQ chatroom while at the same internet café that they both frequented while living in Shenzhen, where she was studying graphic design and he sold computers. Hu Yang used QQ to find potential customers and friends. Lin Ying often went to the internet café to relax and chat with *wangyou* (网友 "net friends"). One day she put out a query to see who was in the same café and Hu Yang replied. They chatted online for a couple of months before their first date (though Hu Yang had already figured out who she was) and eventually got married.

35 Fieldwork in Zhu Village, July 2011.

36 Despite predictions in the 2000s that internet cafés would close due to increased regulations and the diffusion of smartphones and computers, they remained quite popular into the 2010s as sites for gaming (Szablewicz 2020).

37 During this period, the ACWF was promoting the improvement of rural women's lives and their economic empowerment through the use of ICTs (Fan, Xiao, and Li 2007; Tang 2012). These were project-based initiatives and their success (according to the ACWF) contrasted with the findings of scholarly research at the time that found that women, especially those in their 30s and above, were often left out of online informational networks, just as they were in the physical realm (Oreglia 2014; Wallis 2015). Song (2015) documents women who parlayed either their family ties or their migration experience (or both) into what she calls "self-employment as an individual career." Like their male peers, they gained new ideas and skills, yet barriers to their business success included small social networks, a limited ability to find resources, and gender norms that emphasized their responsibility for domestic concerns.

38 The Internet Plus plan, as mentioned in the introduction of this book, emphasizes, among other things, efforts to increase innovation and upgrade various manufacturing capabilities. The 13th Five-Year Plan (2016–2020) echoed the long-term goal of building a "moderately prosperous society" (*xiaokang*) throughout China (Xinhua 2016).

39 When e-commerce first emerged, only small segments of rural China had the infrastructure and knowledge base necessary to carry it out. At the closing of the Third Plenum of the 18th Party Congress in 2013, Xi Jinping issued a 60-point blueprint that emphasized the importance of rural development, agricultural and land ownership reform, and the expansion of information technology in rural

areas. For the original Chinese and an English translation, see https://chinacopy-rightandmedia.wordpress.com. On technosolutionism, see also Lindtner (2020).

40 Xi's 2017 speech was followed by the "Strategic Plan for Rural Revitalization (2018–2022)" (Xinhua 2018), which mapped a detailed plan for transforming the countryside, including the training of "new-style professional farmers" (*xinxing zhiye nongmin* 新型职业农民), who were often returned migrants. In 2019, China's E-Commerce Law also linked rural e-commerce to poverty alleviation (Ministry of Commerce 2019). In November 2020, despite COVID-19, which at the time seemed to be eradicated in China, the government declared victory over poverty, although some observers outside China were skeptical about the figures used in making this declaration (Areddy 2020). In 2021, the government's rural revitaliza-tion efforts were further increased (Donnellon-May 2022).

41 For a detailed account of the interconnection between state policies, market dynamics, and the growth of Alibaba, see Yuan (2021).

42 During my fieldwork in the villages in 2019, livestreaming was mentioned only minimally. However, as I discuss in the book's conclusion, after the COVID-19 outbreak and the intermittent lockdowns, livestreaming, with the help of com-panies like Alibaba, Pinduoduo, and Douyin, became crucial for many farmers and other rural business owners to be able to sell their goods. See Hao (2020) and Tang et al. (2022).

43 Taobao Villages are predominantly located in five eastern coastal provinces: Fujian, Guangdong, Jiangsu, Shandong, and Zhejiang, a pattern that has persisted since they were developed (Aliresearch 2017, 2020b; Qi, Zheng, and Guo 2019, 111). These regions were already more prosperous due to earlier economic policies. Aliresearch frequently mentions model or "Top" Taobao Villages. These include certain towns and villages, including Qingyanliu Village in Yiwu City in Zhejiang, which was one of the first Taobao Villages (A. Li 2017); Suichang County (the so-called "Suichang Model") in Zhejiang (Zi 2019); and Shaji town in Jiangsu (X. Wang 2013). All of these locales have had crucial government support.

44 Ms. Fang's husband also had various jobs. In 2019 he had become an agent for a company that bottled spring water in a nearby village, where he drove each week to collect the water to sell, and he also used his small van as an unofficial local taxi.

45 Tencent and Alibaba are the major players when it comes to mobile payments. In popular discourse, China is seen as basically a cashless society, and this might be true among residents of major cities. However, like so many other phenomena in China, the urban-rural difference is vast.

46 Fieldwork in Suan Village, August 2019.

47 Fieldwork in Suan Village, August 2019. Aside from the online competition, Mr. Shen was selling very few satellite dishes at the time because villagers were streaming television and movies. In 2011, during my first visit, small, semi-legal satellite dishes graced rooftops all over the village. In 2019, there were hardly any to be found.

48 See, for example, this video of Zhang Jingfeng, party secretary of a district in Jinan city in Shandong province, discussing the need for leaders to be transparent and upright: https://dygbjy.12371.cn/2021/03/01/VIDE1614567129633785. shtml#10006-weixin-1-52626-6b3bffd01fdde4900130bc5a2751b6d1. I thank Jie Yang for sending this to me.

49 In contrast, Liu (2020) argues that in the Taobao villages where she conducted research among return migrants who had set up online shops, husbands and wives shared their online bank accounts, and women could therefore exercise a lot of control over family finances.

50 In rural China, a family workshop usually has several nuclear and extended family members working. Again, I am using the term "family business" more broadly, in this case because Ms. Gu and her husband owned the workshop (although she ran it).

51 Fieldwork in Suan Village, August 2019. Ms. Gu's experience of working in a factory that was not extremely far away was typical of the few older women who had had migration experience. Working 80 kilometers away, as opposed to much further away, meant they could go home on weekends or a least once a month to be with their families.

52 The women who sewed at home got a slightly higher piecemeal rate because they were using their own electricity.

53 Like Ms. Pang, Ms. Gu was not in the village WeChat group, but she wasn't in the teachers' group either. She said she had no time.

54 See Khanna et al. (2019) for a detailed account of the development of Rural Taobao through its "1.0" and "2.0" versions.

55 Fieldwork in Suan Village, August 2019.

56 At the time, this would have been roughly $300. Even in the village, this amount was not considered a lot of money, especially given all the time Ms. Jing spent.

57 Her experience was somewhat similar to a story related by MacDonald (2016, 171) about a woman who opened her own online Taobao store but sold clothes through it as an agent for another seller. She eventually closed the store because it was too much time for too little money.

58 Yu and Cai (2109) state that, although younger women with some migration experience gained economic empowerment and a voice through e-commerce, they did not achieve political or cultural empowerment, a phenomenon they call the "femininization of Taobao villages" (430). These women continue to bear the brunt of household labor, doing the "inside work." Their findings correspond with my earlier work on ICTs and micro-entrepreneurship in rural China (Wallis 2015).

59 Other research has also found some minimal use of WeChat for rural micro-entrepreneurship. Wang and Sandner (2019) conducted research in a township (not a village) outside Changsha and found that a small number of older married women who had never migrated had opened their own businesses. They used WeChat to enhance sales from their physical stores (e.g., through showing photos

of new products/arrivals), yet they were not engaging in what would be considered e-commerce.

60 Fieldwork in Suan Village, August 2019.

61 As discussed in the book's introduction, WeChat has various functions through which users can attempt to add strangers as contacts.

62 Fieldwork in Suan Village, August 2019.

63 The blurring of commerce and friendships through this use of WeChat, not surprisingly, became a societal concern at the time (Wang 2015).

64 There are numerous articles written online that offer advice on how to successfully use WeChat to sell things. I thank Haoming Zhou for sending me the following example: https://zhuanlan.zhihu.com/p/67647738.

65 WeChat's shake function allows users to literally shake their phone and connect with nearby users who have this function turned on.

66 Danchen Liu (personal correspondence). I also thank Danchen for reminding me of the village gossip that such schemes can generate.

67 As Low (2018) reports, the vast number of people who were involved with *Shanxinhui*, combined with the fact that several of them participated in a large demonstration in Beijing after the founder's arrest, prompted the government to launch a crackdown on such schemes.

68 Pyramid schemes have been a problem in China for years, but their appeal and volume of investment seems to have grown since the early to mid-2010s. See www.economist.com.

3. DOMESTIC WORKERS, VIRTUAL VOICE, CARE, AND COMMUNITY

1 Fieldwork in Beijing, October 2017. All names are pseudonyms. On domestic workers' consciousness of their subordination, see Ellerman (2017).

2 Part of the discussion also centered on lax safety measures at residential compounds that have been built in recent years to accommodate China's uber-wealthy. Firefighters arrived only to find that there was not enough water pressure for them to douse the blaze on the 18th floor apartment (Koetse 2017). The surviving husband eventually filed a lawsuit. See www.sohu.com.

3 Fieldwork in Beijing, October 2017. When this topic came up in a private conversation with a domestic worker in 2019, she expressed the same sentiment.

4 The post was written by a domestic service agency and can be viewed at https://mp.weixin.qq.com/s/GXu-wmmgEIzdDGaPw-RX_g.

5 Many urban residents express a distrust of domestic workers (iiMedia Life and Travel Research Center 2022).

6 Of course, some domestic workers feel they are treated fairly well, and many have been with the same employer for several years. Still, even in the best of circumstances, many domestic workers can experience discrimination and alienation.

7 This chapter incorporates and builds on previous fieldwork (see Wallis 2018, 2022). I base my arguments on extended fieldwork conducted in Beijing during 2013, and follow-up visits in 2014, 2015, 2017, and 2019, as well as digital ethnog-

raphy. The participants were 25 female domestic workers who ranged in age from 35 to 55 and who had been in Beijing anywhere from two to twenty-one years. I conducted semi-structured interviews with them (in some cases in pairs) and focus groups and also engaged in casual conversations with many of them on multiple occasions. Most of the women were from rural areas, except three who were laid-off factory workers from large cities in central and western China. I met 14 of the women in 2013, two more in 2014, and the other nine in 2015 and 2017 through attending workshops and other activities hosted by different NGOs that serve the needs of domestic workers in Beijing. I conducted extensive participant observation among a subset of the women who were involved in an NGO-organized drama club that met all day once a week. In addition to attending the weekly workshops several times over the years, I participated in various social activities and viewed their performances, which enabled me to form bonds with a core group of women. I also interviewed the drama club organizer (a college-educated NGO staff member). At three other NGOs, I also attended workshops and other activities, and I interviewed the co-founder of one of the NGOs. The chapter is also based on analysis of hundreds of posts and photographs of several of the women's Qzone (QQ*kongjian*) and WeChat for common themes, off and on during this time period. Analyses of their posts are the result of in-person sharing, chats via social media, and intermittent digital ethnography. Some women only posted occasionally, but others were more prolific. At the beginning, all the women except one used QQ in various ways—for chatting, listening to music, reading news, and social networking. Over time, they all shifted to WeChat as their primary social media platform. This shift happened in tandem with all of them acquiring low-end smartphones with internet access. Finally, I was a member of the drama club WeChat group for a number of years, but I was invited into the group as a "sister," not as a researcher, so I do not include any information on what was shared in that group in this chapter.

8 Such fictive kin terms are commonly used in China and are meant to communicate a warm relationship and/or affinity/familiarity.

9 Both groups had basically been left to fend for themselves, although those who were laid off from state-owned enterprises received a pension. According to my informants, however, it was too small for them to live on. The laid-off workers' status as domestic workers was especially humiliating for them because, prior to the reforms, they were part of the prized urban working class who were guaranteed an "iron rice bowl" (cradle-to-grave social welfare benefits, including housing, healthcare, education for children, etc.). See also Yang (2015).

10 Yan (2008, 129) further argues, "The labor mobility of migrants is coded as a form of accumulation of *suzhi* that will lead to self-development (*ziwo fazhan*) and class mobility. What this coding makes invisible is how the labor power of migrant workers creates surplus value for capital accumulation."

11 Fieldwork in Beijing, September 2015.

12 On NGOs, "voice," and labor migrants more generally, see Gleiss (2014).

13 COVID-19 and the stringent Zero-COVID policy slowed this growth temporarily, but it was expected to recover (Mo, Zhang, Ding, and Yang 2022).

14 Maternal care means taking care of a new mother and her newborn child for 30 days after the child is born. See also Ding (2023).

15 See the introduction for a discussion of the Internet Plus plan.

16 Fieldwork in Beijing, September 2015. This woman knew another woman who had spent 8,000 *yuan* on trainings from the same company.

17 Even worse exploitation happens when a domestic worker is coerced by an employment agency into going to a new employer's home before signing a contract. One domestic worker told me that the contract she signed omitted that she was expected to take care of an infant full-time. The language in the contract only said, "help with." To make matters worse, the employer didn't want her to use gloves when changing the infant's diaper.

18 See also www.gov.cn/zhengce/2019-07/06/content_5406675.htm.

19 According to Hochschild (2003, 7), in contrast to the feelings and often unspoken negotiated rules that guide private "emotion work" or "emotion management," "which has *use value*," emotional labor is "sold for a wage and therefore has *exchange value*" (italics in original). It is a type of commodity that is institutionalized and standardized through various mechanisms, including company rules and training sessions. Hochschild adds that achieving the transference of private acts of feeling into commodified emotional labor is done through forms of social engineering that can result in either surface acting, or feigning a display of feeling, or deep acting, where one alters herself to have the desired feeling (here Hochschild is using the Russian director Constantin Stanislavski's distinctions). Over time, as outward display does not always match inward feeling, "emotive dissonance" can lead to stress and self-alienation (90).

20 Scholars researching domestic workers in various contexts have found Hochschild's (2003) concept of emotional labor useful, even though domestic workers are in private homes (see Hondagneu-Sotelo 2001; Huang and Yeoh 2007; Ma 2010). For an extended discussion, comparison, and critique of Hochschild's emotional labor and Hardt's (1999) affective labor, see Weeks (2007).

21 Fieldwork in Beijing, April 2013.

22 Fieldwork in Beijing, October 2017. See also Gaetano's (2004, 55) discussion of domestic workers and "virtual servitude."

23 Some domestic workers are also the targets of physical violence, although none of my informants had this experience as far as I'm aware.

24 Fieldwork in Beijing, April 2013. In all of my fieldwork trips, domestic workers shared these types of experiences. In 2017, while I was in Beijing, another woman told a similar story of being blamed for minor scrapes on a child's arm. For more on such humiliations, see Gaetano (2015).

25 As discussed in the introduction, ever since the government's numerous crackdowns in 2021, celebrities and influencers are less likely to flaunt their wealth.

26 For further discussion of Ahmed's ideas, see introduction, endnote 33.

27 Although Qzone and WeChat are different platforms, I combine my analyses of women's posts on the two platforms because the content of their posts did not change once they switched from Qzone to WeChat as the primary platform they used. The main difference between the two is that Qzone users can see all comments on a friend's post even if they are not friends with the person who posted the comment. On WeChat, users can only see comments by those who are also their WeChat friends (or contacts). This difference did not present an obstacle because I was friends with many of the women on WeChat so I could see their comments to each other. By 2015, all the domestic workers in this chapter were primarily using WeChat.

28 This work is indebted to Rakow and Navarro's (1993) early research on "remote mothering" via mobile phones.

29 Fieldwork, September 19, 2015. Women forwarded, as opposed to wrote, content for a number of reasons, including varying levels of active (as opposed to passive) literacy, time constraints, convenience, and ideas regarding what was interesting or worthy of posting. The forwarded posts usually appeared with no or minimal comments added. Forwarded posts were the most common type of post. Also, some women did post "hard news," especially during the "migrant clean-up" in 2017 in Beijing, when the government expelled vast numbers of the so-called "low-end population" (see the introduction), and after COVID-19 emerged, but these posts were far fewer than the types I discuss here.

30 Admittedly, many of these could have been misinformation.

31 See the introduction and chapter 1.

32 Such sentiments have popularly been called "chicken soup for the soul."

33 As a contrast, see Peng (2019) for an insightful analysis of how young urban professionals display pictures of food (often that they have ordered at restaurants) on WeChat to construct a middle-class aesthetic.

34 Fieldwork in Beijing, October 2017.

35 Fieldwork in Beijing, October 2017.

36 In 2020, during the first wave of COVID-19, a few women sent me marketing messages regarding different business opportunities, although it wasn't always clear if they were actually involved in these promotions or were merely spreading the word.

37 Before the women had smartphones, such posting could take quite a bit of effort. In the past, Ms. Li, for example, did not have a camera phone or a computer, so in order to post photos on her Qzone she went to a photography shop and had the staff transfer the photos first from her camera to the store's computer, and then from the computer to her Qzone. By 2015, this was not an issue because all of the women had smartphones.

38 On the imaging practices of younger migrant women, see Wallis (2013a).

39 Fieldwork in Beijing, June 2014. Except for several core members, membership in the club was relatively fluid.

40 As mentioned earlier, WeChat users can put contacts into different groups. This practice is very common and aligns with the way that Chinese relationships are

defined by particular experiences and connections (classmates, colleagues, hometown people, etc.).

41 Fieldwork in Beijing, October 2017.

42 Fieldwork in Beijing, August 2019.

4. A NETWORKED FEMINIST KILLJOY ASSEMBLAGE

1 The Chinese title translates literally as "The Way of Vagina," but the English translation references the classic 1970 book *Our Bodies, Ourselves* that emerged out of the feminist women's health movement in the United States.

2 Bcome is a symbolic name. It sounds like "become," which speaks to potentiality, a hallmark of affect. "*Bi*" is also a Chinese slang word for a woman's vagina; followed by "come" the meaning is quite obvious, implying women's pleasure. For more on Bcome, *The Vagina Monologues*, and *Yiyuan gongshe*, see Wang (2018a).

3 To me, the frank language used to address the various issues covered was surprising and refreshing. Previously, I had only heard some of the words hurled as insults during arguments on the streets of Beijing, or spoken among small groups of women friends sharing personal experiences behind closed doors.

4 For a more detailed analysis of this event and the reaction to such performances in general, see Huang (2016). The various performances in 2013 marked the 10-year anniversary of feminist scholar/activist Ai Xiaoming's introduction of *The Vagina Monologues* to her students. See also Wu and Dong (2019).

5 Other early actions included the "Wounded Bride," in which three young women adorned in blood-stained wedding dresses strolled through a popular tourist area in Beijing to raise awareness about domestic violence. See, for example, Fincher (2018), Li and Li (2017), Tan (2023), Wang (2018a), Wei (2015), and Behind the Wall (2012). Even Chinese state media initially reported on these events favorably (see Li 2012). In English language scholarship, this loose network is also called the Youth Rights Feminist Action School (Tan 2023) and Youth-Feminist-Action-Faction (Wang 2018a).

6 The statement regarding 2012 as the "Year of Feminism" in China contains some hyperbole and erasures. It ignores earlier efforts by feminists, such as those protesting the arrest in 2009 of Deng Yujiao, a restaurant worker who stabbed and killed a CCP official after he tried to sexually assault her. Many believe that public outrage online led to the charges against her being dropped and to her release. See Fincher (2018, 24–25).

7 In March 2015, the five women, later known as the "Feminist Five," were arrested and detained for planning to hand out stickers to raise awareness about sexual harassment on public transportation, as mentioned in the introduction. See Branigan (2015), Tan (2023), and Fincher (2018).

8 See www.sohu.com/a/141789155_114731. The ACWF is one of the country's eight original "mass organizations" that receive guaranteed funding and support from the government.

9 During the spring and summer of 2021, several WeChat LGBTQ student accounts were deleted (The China Team 2021). Also, high-profile activists, including Xiao Meili and Liang Xiaowen, were viciously trolled by extreme nationalists, or "Red Vs," as they are called, who often work in concert with the state, and their Weibo accounts were deleted as well (Gan 2021).

10 "Equality between men and women" is enshrined in law and has been a founding principle of the Chinese Communist Party since its earliest days (see Wang 2017).

11 I discuss these historical linkages in the next section, but I want to point out here that "Occupy Men's Toilets" was inspired by both an earlier action by female college students in Taiwan and the Occupy Wall Street movement in the United States in 2011. The "Wounded Bride" was based on an action in Turkey in 2011 (Wang 2012).

12 In Russia, for example, feminists are attacked for not adhering to the notions of conventional gender and sexuality that are part of Vladmir Putin's nationalist agenda (Surman and Rossman 2022).

13 Of course, such contemporary reactions are not completely new. US feminists of the late 1960s and 1970s were derided as man-hating, hairy, ugly radicals who wanted to destroy American families and society. Banet-Weiser's (2018, 1) argument regarding popular misogyny is that it is a corollary to popular feminism in the US and elsewhere. By popular, she means that its "discourses and practices . . . are circulated in popular and commercial media," it is "liked or admired by like-minded people and groups," and, referencing Stuart Hall, it is a site of cultural struggle.

14 As discussed in the book's introduction, the promotion of extremely conservative gender ideologies has been a hallmark of Xi Jinping's rule. As the economy continued to suffer post-Zero-COVID and the demographic crisis loomed large, Xi Jinping and state rhetoric were more explicit about women's proper role in the home (see Stevenson 2023). The current call for women to be in the home has historical precedent in various "return home" debates over the decades (see Song 2016). Of course, the contemporary state rhetoric seems particularly anachronist to many young women who have a stronger sense of their individual desires and who are influenced by transnational discourses regarding women's independence and personal choice.

15 This chapter is based on fieldwork conducted in Beijing during different periods of time between 2015 and 2019, interviews, and textual analysis of a range of content delivered through various social media platforms. Semi-structured interviews, many with follow-up conversations, were conducted with 19 young women who identified as feminists. They ranged in age from 19 to 32, with most in their twenties. Most were college graduates, although three were still undergraduates. Many spoke English and some had studied abroad to obtain a master's degree. Interviews were conducted in person or over WeChat. During the longer fieldwork for the book, I attended various events focused on gender and sexuality, such as OVO, film festivals, NGO workshops, and other gatherings. I am also a member of several

feminist WeChat groups, although I do not include details of those groups here in order to protect the privacy of group members. I also do not include any analysis of informants' personal social media posts for the same reason, given the sensitive nature of feminism in China, and I provide general descriptions rather than exact translations of content posted on public platforms. Finally, the groundwork for this research was laid over many years through my friendship and association with several older feminists who are "in the system" in China.

16 Ngai (2005) invents the word "stuplimity" to describe a combination of shock and boredom, but I do not adopt that term.

17 In formulating this term, Ito (2008, 3) states that "publics" was chosen "to foreground how people are reactors and (re)makers in relation to media, engaging in shared culture and knowledge through discourse and social exchange as well as through acts of media 'reception.'" boyd (2010) adds that networked publics enable an "imagined collective" (39) that has "been transformed by networked media" (42). Papacharissi (2015) builds on this formulation in her analysis of "affective publics," which centers the role of affect and emotion in broad-based social movements, such as the Arab Spring and Occupy Wall Street. She draws on big data analysis of anonymous social media users. See also Garde-Hansen and Gorton (2013).

18 China's #MeToo movement officially started when Luo Xixi, a Chinese graduate student studying in the US, posted an open letter on Weibo, accusing a former professor in Beijing of sexually assaulting her. The movement then spread through Chinese universities before it diffused through China's business, entertainment, and political domains.

19 See endnote 9 above. Lynette Ong (2022) discusses what she calls the "outsourcing of repression," such as the use of violent "thugs-for-hire," in the context of land grabs in rural areas as part of China's urbanization drive.

20 These complementary events were preceded by numerous planning meetings and study sessions for attendees, who came from all over China, so that they would be prepared to participate. Although the word "feminist" was not used to frame the conference, two Chinese friends who are now prominent feminist scholars told me that participating in the activities before, during, and after the conference was transformational in their thinking about women's and gender issues and becoming feminists.

21 See www.un.org/womenwatch.

22 See, for example, Zhu and Xiao (2021), Wu and Lansdowne (2019) "Introduction," Fincher (2018, chap. 6), and Tan (2023). Many of the journal articles on the YFAs that I cite in this chapter also include briefer accounts.

23 The most well-known Confucian reformers who advocated for women were Liang Qichao and Kang Youwei. The May 4th Movement was sparked by student protests on May 4, 1919, against what the students thought was China's weakness in the face of imperialism. The New Culture Movement was a broader social/intellectual movement in the 1910s and 1920s that sought to throw off Confucianism

and modernize China through embracing science and individual rights, among other ideas. The three obediences are: a woman should obey her father before marriage, her husband while married, and her eldest son when she is a widow. The four virtues regulate women's speech, appearance, conduct, and work. See Du and Cai (2005) for a history of Confucian patriarchal gender relations. Edwards (2000) notes that, as the "woman question" became articulated to national salvation, the modern woman became a key figure in literature and popular culture, and eventually a commercial signifier in magazines and advertisements as well. On gender and May 4th/New Culture reforms, see also S-L. Zhang (2007).

24 These words entered China most often through Chinese translations of Japanese translations of western works.

25 Yenna Wu (2005, 47) references Chen Dongyuan (1927/1981) on other terms for feminism during this time: "'Feminism' was either transcribed as *fominieshimu*, or translated as *nannü pingquan zhuyi* (the -ism of equal rights/power of men and women) or *funü zhuyi* (the -ism of women)" (endnote 4). See also Sudo (2007) for a detailed discussion. Feudalism was a catch-all term for the perceived backward gender norms prescribed by Confucianism (Wang 2017, 3).

26 The CCP was founded in 1921. For CCP cadres, the writings of Frederick Engels on women and the family served as their ideological foundation (Jacka 1997; Wang 2017). Some urban intellectuals and feminists within the early CCP were fully committed to women's emancipation at this time, so when *nüquanzhuyi* was rejected, according to Wang Zheng (2017), they pushed instead for the adoption of other terminology, such as "women's rights" and later, "women's work." Achieving buy-in from other male cadres and male peasants (China was largely agrarian) was a struggle. In her book, Wang documents how, in the early days of the CCP and throughout this period, a group of state feminists toiled behind the scenes to fight for women's equality in a deeply patriarchal society.

27 Equality between men and women was included in the 1954 Chinese Constitution. The Marriage Law (1950) abolished arranged and child marriages and granted women the right to divorce. However, in practice the law was implemented unevenly, especially in rural areas (Wu 2005).

28 The Iron Girls were strong, technically skilled, patriotic women who were featured in Party propaganda doing a range of "masculine" jobs (fixing power poles, operating heavy machinery, etc.). See Wang (2017, 222–228).

29 The ACWF was the first to proclaim itself an NGO in 1993. While many initially balked at this designation, over time, many women's groups and international agencies accepted the ACWF as an NGO (Liu 2001). More recently, the term GONGO has been used. The alarming phenomena that all these groups were trying to combat included employment and education discrimination, domestic violence, and, in rural areas, trafficking in women and female infanticide, the result of the one-child policy established in 1979.

30 At the time, Li Xiaojiang was a professor at Zhengzhou University. She was a vocal opponent of both the ACWF and western liberal feminism. For a comparison

of the ideas of Li Xiaojiang and Du Fangqin, another founder of women's studies in China, see Spakowski (2011).

31 When literary scholar Zhang Jingyuan (1992) edited and translated into Chinese a collection of western feminist writings, she wrote that she chose to use *nüxingzhuyi* as the translation of feminism because she thought it was the best term to use for people to accept feminism. To her, "*nüquanzhuyi*," which highlights women's rights and power, would be too associated with *nüqiangren* (女强人, or strong woman) and the "equality between men and women" discourse, so maligned in the wake of the repudiation of the Mao era. Also, *quan* and *li*, meaning "rights" and "power," might be understood as women taking rights away from men. *Nüxingzhuyi* has continued to be used by some women's studies scholars, but others reject western feminism completely for its connection to capitalism and as not suitable for China. For example, Huang Lin, a scholar who founded (in 2004) and edits the journal *Feminism in China* (*Zhongguo nüquanzhuyi* 中国女性主义) has put forward the idea of "smiling feminism" to emphasize harmony between men and women. In the first issue of the journal (given to me by one of the contributors), the front leaf says that Chinese feminism (*Zhongguo nüxingzhuyi*) is "not aggressive," focuses on "harmonious development," and represents a "smiling feminism" (*wei women biaoshichu weixiaozhede zhongguo nüxingzhuyi*). For a discussion of academic debates on feminism, see Spakowski (2018, 2021).

32 Of course, the state/civil society divide does not exist in China in the way it does in the West. For a deeper analysis of this phenomenon in China, see Wang and Zhang (2010).

33 For some, not identifying as "feminist" is a strategic choice, i.e., they can accomplish their goals without drawing unwanted attention. Others still view their work through the pre-reform lens of "women's work." Still others believe feminism is divisive, as mentioned above. For a discussion of the differences between the "feminism" practiced among women's and migrant worker NGOs, consumerist empowerment feminism, and activist feminism, see Yin (2022).

34 I have a vivid memory of visiting a Chinese friend, who is a feminist writer and professor, shortly after the e-paper launched. She was extremely excited about its content, and she particularly praised Lü Pin's sharp analysis and original thought. Lü Pin eventually became one of the most well-known feminist activists in China and was closely linked to the YFAs. Since the detention of the Feminist Five, she has resided in New York and carries out her activism there. See the China Change interview with Lü Pin discussing her motivation to try to make feminism relevant to society and social change: https://chinachange.org. For a more detailed history of Feminist Voices, see Han (2018) and Wang and Driscoll (2019).

35 On Sora Aoi, see https://asiatimes.com. On Liu Jishou, see https://chinasmack.com.

36 Here and elsewhere I note the year they were born rather than their age because I interviewed them at different times. I don't include where Hai got her master's degree to maintain her privacy.

37 See Fincher (2018), Wu and Dong (2019), and Wang (2021). Wang Qi (2018a) in particular provides a compelling account of how shared experiences as an age and historical cohort informed the feminist praxis of several high-profile YFAs.

38 Fieldwork in Beijing, August 2019.

39 WeChat interview, May 2018.

40 Fieldwork in Beijing, 2019.

41 WeChat interview, May 2018.

42 See the introduction, endnote 22.

43 Fieldwork in Beijing, August 2019.

44 Post COVID, many young Chinese women, not just feminists, are rejecting marriage.

45 Now, individual self-interest in the material realm is more acceptable, and in many ways encouraged, but it is not in interpersonal relationships.

46 Fieldwork in Beijing, August 2019.

47 Several terms for feminism, most of them derogatory, have emerged in popular culture. For example, *tianyuannüquan* (田园女权), translated variously as "pastoral," "grassroots," "rural," or "localized" feminism, became popular in the mid-2010s. It is a derisive term for women who follow a form of feminism that could best be described as a sort of selfish, market-oriented feminism. Some feminist activists, such as Lü Pin and Xiao Meili, have tried to appropriate the term to diffuse its power.

48 WeChat interview, April 2020. She also mentioned reading *How the Steel was Tempered* (*Gangtie shi zenyang lianchengde*), a 1934 socialist realist novel by Nikolai Ostrovsky about the Russian revolution. As she stated, "There was a very famous quote from that novel, that people should not waste their time or have regrets for what they do in their life. . . . I was influenced by that because I felt like people should—it's not money, or like just a good life I was looking for . . . I was looking for this idea for life."

49 See note 47 above. Another term, "Weibo Feminism," is used to denote more radical feminists who, among other things, oppose marriage and surrogacy and seek to revive forgotten women's history (Xue and Rose 2022).

50 Fieldwork in Beijing, October 2017.

51 WeChat interview, July 2018.

52 See Wallis and Shen (2018).

53 Some scholars suggest that China is "postfeminist," yet the term needs to be qualified in the Chinese context. It is true that the beauty standards and emphasis on individualism, empowerment through the body and through consumption, the hyper-feminine appearance presented in commercial media, and notions of natural binary gender fit with what Rosalind Gill (2007) calls a "post-feminist sensibility," which is also visible globally (Dosekun 2015). However, I would argue that these have not emerged as a result of Chinese women "taking feminism into account" for what it has brought them and then repudiating it as no longer necessary, which, according to Angela McRobbie (2004), is a hallmark of

postfeminism. Rather, the gender milieu in China is an assemblage of Confucian (what Mao called "feudal," which is the term May Fourth intellectuals used) ideas regarding binary gender; the legacy of Mao-era policy and rhetoric of "equality between men and women" and "women's work" (called "state feminism" in scholarship but not a term used at the time); the particular neoliberal tenets China has embraced in the social realm, namely an emphasis on individuals rather than structures and intense marketization and commodification of everyday life (and women's bodies); global flows of images and ideas; and the demographic imbalance caused by the former one-child policy. Similarly, "neoliberal feminism" (Rottenberg 2018) does not describe gender dynamics in China, either, because China is not neoliberal in the western sense of the word. Wu and Dong (2019) call this "entrepreneurial feminism." See also Yin (2022).

54 I thank Dai Qingyu for sending me these memes. 96 *jin* is roughly 106 pounds.

55 WeChat interview, May 2018.

56 *Lean In* was very popular in China, where it was translated into Chinese. In cities, there were reading groups to discuss and dissect the message of the book. See Tatlow (2013). It was also widely critiqued in and outside of China for Sandberg's simultaneous acknowledgement of and disregard for social structures and her espousal of what Rottenberg (2018) calls neoliberal feminism, which is based on choice, market rationalities, and individualized notions of empowerment. One young feminist, Jia, had participated in a *Lean In* group in Beijing as a volunteer. She said she thought Sheryl Sandberg was a feminist but that her Chinese fans were not. Of the other participants she said, "They really cannot agree with feminism. They argued a lot." WeChat interview, September 2018.

57 WeChat interview, July 2018.

58 WeChat interview, September 2018.

59 WeChat interview, May 2018.

60 WeChat interview, September 2018.

61 WeChat interview, September 2018.

62 WeChat Interview, July 2018.

63 WeChat Interview, July 2018.

64 Fieldwork in Beijing, August 2019. In recent years, those who are antagonistic to feminism replace "*quan*" (rights and power) with the character for "fist" (they are homonyms in Chinese) to chastise feminists for being too aggressive.

65 As a result of Occupy Men's Toilets, there were changes in the ratio of men's to women's public toilets in Guangzhou (Wei 2015). After the YFAs started a campaign to protest unfair college admissions based on gender, in May 2013 the Ministry of Education prohibited higher education institutions from setting gender quotas for enrollment (Tan 2023). The YFAs also joined forces with older feminists and women's studies scholars, who had long fought for an anti-domestic violence law, to push for the law, which was passed in March 2016 (Fincher 2018).

66 Independent feminist activism in general has also been critiqued for being too focused on the concerns of urban, educated, middle-class women even though

many young activists have also dedicated their time and energy to social causes, such as volunteer work among domestic workers and labor rights.

67 Fieldwork in Beijing, August 2019. All quotes that follow in this section are from this same conversation.

68 See Wu and Dong (2019) on what they call "entrepreneurial Chinese feminism."

69 Livestreaming is ephemeral, and Fang Ai never archived her livestreams, so I was not able to view any of them.

70 For a useful primer comparing YouTube to Bilibili, see Greater China Tech's "How China's Video Platform Bilibili Works," www.youtube.com/watch?v=A200wcPatV4. See also Craig, Lin, and Cunningham (2021).

71 Over time even private groups became subject to stricter surveillance and had to self-censor. Groups are often shuttered by the platforms and then started as new accounts by the group moderators. I know this from my own experiences of being in such groups.

72 The show chronicles the lives of two young women who want to start a cupcake business. It was hugely popular, including among some of the young women featured in this chapter, and among college students and young white-collar workers (who did not necessarily identify as feminist) that I knew in Beijing while doing fieldwork for this book. Its popularity presumably lies in these viewers' identification with the struggles (with boyfriends, finances, trying to pursue a dream, etc.) of the two protagonists.

73 I do not have space to discuss more examples, but one serious video that was translated was from Jubilee, whose motto is, "We exist to create a movement of empathy for human good" (www.jubileemedia.com/vision). It was from Jubilee's "Seeking Secrets" series (https://imari.typeform.com), in which people were asked to send in their deepest darkest secrets and have strangers read them.

74 The YouTube video can be viewed here: www.youtube.com/watch?v=VvKmgTpBffo.

75 The fansubbed clip had 3,800 views and 12 comments, all positive, when I watched it. The original *Ellen* clip has received over 1.5 million views and over 1,300 comments. Yet, for the fansub group, the point was not reaching millions of viewers; it was making a feminist statement.

76 WeChat interview, May 2018.

77 The connection between "impression management" and digital media has been well-established since the earliest research on mobile phones and what (at the time) were called social networking sites. See Fortunati (2005) and Ellison, Heino, and Gibbs (2006).

78 WeChat interview, September 2018.

79 WeChat interview, May 2018.

80 WeChat interview, June 2018.

81 WeChat interview, May 2018.

82 WeChat, like Facebook, allows users to block certain people and make a post visible only to selected people. It also has a Time Limit setting that allows users to

make all posts visible or to only show posts from the last three days, last 30 days, or last six months. See Huang, Vitak, and Tausczik (2020, 468) for an explanation of the various setting options.

83 WeChat interview, September 2018.

84 WeChat interview, May 2018.

85 See Wu and Dong (2019) for a thoughtful critique of this trend.

86 Imperfect Victim can be viewed with English subtitles here: www.iq.com.

CONCLUSION

1 The test had actually happened in February, but the news was not released until later.

2 On the Wuhan lockdown, see Yang (2022). On the range of emotions expressed on WeChat public accounts during the first iteration of the pandemic in China, see de Kloet, Lin, and Hu (2021).

3 Both of these events sparked outrage and also renewed debates about gender violence in China. See Kuo (2022) and Ni (2022b).

4 This trend is predicted to increase the demand for domestic workers. See "China's Largest Online Domestic Helper Agency" (2023).

5 See https://chinadigitaltimes.net.

6 On my host's WeChat Moments, there were pictures that showed continuous transformations—more paved roads, streetlights, garbage dumpsters, etc. in the village as well.

BIBLIOGRAPHY

Ahmed, Sara. *The Promise of Happiness*. Duke University Press, 2010.

———. *The Cultural Politics of Emotion*. 2nd ed. Edinburgh University Press, 2014.

Aliresearch. "Zhongguo Taobaocun Yanjiu Baogao" [China Taobao Village Research Report 2014]. 2014. www.aliresearch.com.

———. "Make a Better Rural China." June 14, 2016. https://data.alibabagroup.com.

———. "Ali Fabu <Nüxing Chuangyi Shehui Zeren Dashuju> Mama Zhubo Xi-eqi 'Tuixiu bu tuichao' de Chuangye Jufeng" [Ali Released "Big Data on Social Responsibility for Female Entrepreneurship" Moms become Popular Streamers as Entrepreneurs]. March 8, 2019. www.aliresearch.com.

———. "'Nongcun Taobao' Jinhuashi: Hulianwang Ruhe Gaibian Xiangcun" [The Evolutionary History of "Rural Taobao": How the Internet has Changed Rural China]. July 8, 2020a. www.aliresearch.com.

———. "2020 Zhongguo Taobaocun Yanjiu Baogao: 1% de Gaibian, 1 anyi GMV | Baogao" [2020 China Taobao Village Research Report: 1% Change, 1 trillion GMV | Report]. October 10, 2020b. www.aliresearch.com.

———. "Nüxing Jiuye Chuangye Yanjiu Baogao: Shuzi Jingji Xianzhu Suoxiao Xingbie Chayi" [Research Report on Women's Employment and Entrepreneurship: Digital Economy Significantly Narrows Gender Gap]. March 7, 2022. www.aliresearch.com.

Alpermann, Björn. "China's Rural-Urban Transformation: New Forms of Inclusion and Exclusion." *Journal of Current Chinese Affairs* 49, no. 3 (2020): 259–268.

Ames, Roger T. *Confucian Role Ethics: A Vocabulary*. Hong Kong: The Chinese University of Hong Kong Press, 2011.

Ames, Roger T., and David L. Hall. *Dao De Jing: Making this Life Significant. A Philosophical Translation*. Translated and with commentary by Roger T. Ames and David L. Hall. New York: Ballantine Books, 2003.

Anagnost, Ann. "The Corporeal Politics of Quality (*Suzhi*)." *Public Culture* 16, no. 2 (2004): 189–208.

Areddy, James T. "China Says it has Met its Deadline of Eliminating Poverty." *Wall Street Journal*, November 23, 2020. www.wsj.com.

Arnold, Frances. "The World's Art Factory is in Jeopardy." Artsy. June 22, 2017. www.artsy.net.

Bakhtin, Mikhail Mikhaïlovich. *The Dialogic Imagination: Four Essays*. Edited and translated by Michael Holquist. Edited by Caryl Emerson. Austin: University of Texas Press, 1981.

Bamman, David, Brendan O'Connor, and Noah Smith. "Censorship and Deletion Practices in Chinese Social Media." *First Monday* 17, no. 3 (2012).

Bandurski, David. "Shifang Protests: 'Permission Denied.'" China Media Project, July 4, 2012. http://chinamediaproject.org.

———. "'Positive Energy,' a Pop Propaganda Term?" China Media Project, November 12, 2014. http://chinamediaproject.org.

Banet-Weiser, Sarah. *Empowered: Popular Feminism and Popular Misogyny*. Durham, NC: Duke University Press, 2018.

Barlow, Tani. "Theorizing Woman: *Funü, Guojia, Jiating* [Chinese Women, Chinese State, Chinese Family]." In *Feminism and History*, edited by Joan Wallach Scott, 48–75. Oxford: Oxford University Press, 1996.

Barry, Thomas. "Affectivity in Classical Confucian Tradition." In *Confucian Spirituality*, Vol. 1, edited by Weiming Tu and Mary Evelyn Tucker, 96–112. New York: Crossroad Publishing, 2003.

Bauman, Zygmunt, and David Lyon. *Liquid Surveillance: A Conversation*. Malden, MA: Polity, 2013.

Behind the Wall. "'Occupy Toilets' Seeks to Double Potty Parity for Chinese Women." NBC News, February 25, 2012. www.nbcnews.com.

Benney, Jonathan, and Jian Xu. "The Decline of Sina Weibo: A Technological, Political, and Market Analysis." In *Chinese Social Media: Social, Cultural, and Political Implications*, edited by Mike Kent, Katie Ellis, and Jian Xu, 221–236. New York: Routledge, 2017.

Berlant, Lauren. *Cruel Optimism*. Durham, NC: Duke University Press, 2011.

Bian, Yanjie. "The Prevalence and the Increasing Significance of *Guanxi*." *The China Quarterly* 235 (2018): 597–621.

Blackman, Lisa. *Immaterial Bodies: Affect, Embodiment, Mediation*. Sage, 2012.

Boeri, Natascia. "Challenging the Gendered Entrepreneurial Subject: Gender, Development, and the Informal Economy in India." *Gender & Society* 32, no. 2 (2018): 157–179.

Bondes, Maria, and Günter Schucher. "Derailed Emotions: The Transformation of Claims and Targets during the Wenzhou Online Incident." *Information, Communication & Society* 17, no. 1 (2014): 45–65.

Bossen, Laurel. *Chinese Women and Rural Development: Sixty Years of Change in Lu Village, Yunnan*. Lanham, MD: Rowman & Littlefield, 2002.

Boullenois, Camille. "Poverty Alleviation in China: The Rise of State-Sponsored Corporate Paternalism." *China Perspectives* no. 3 (2020): 47–56.

Bourdieu, Pierre. *Distinction: A Social Critique of the Judgement of Taste*. Translated by R. Nice. Cambridge, MA: Harvard University Press, 1984.

boyd, danah. "Social Network Sites as Networked Publics: Affordances, Dynamics, and Implications." In *A Networked Self: Identity, Community, and Culture on Social Network Sites*, edited by Zizi Papacharissi, 39–58. New York: Routledge, 2010.

Branigan, Tania. "Five Chinese Feminists Held over International Women's Day Plans." *The Guardian*, March 12, 2015. www.theguardian.com.

Bray, David, and Elaine Jeffreys, eds. *New Mentalities of Government in China*. New York: Routledge, 2016.

Bray, Francesca. *Technology and Gender: Fabrics of Power in Late Imperial China*. Berkeley: University of California Press, 1997.

Brewster, Thomas. "The U.S. Just Charged Huawei with Stealing a T-Mobile Robot Idea." *Forbes*, January 28, 2019. www.forbes.com.

Brown, Wendy. *Undoing the Demos: Neoliberalism's Stealth Revolution*. Cambridge, MA: MIT Press, 2015.

Brussevich, Mariya, Era Dabla-Norris, and Bin Grace Li. "China's Rebalancing and Gender Inequality." IMF Working Paper. May 11, 2021.

Bucher, Taina, and Anne Helmond. "The Affordances of Social Media Platforms." In *The SAGE Handbook of Social Media*, edited by Jean Burgess, Alice Marwick, and Thomas Poell, 233–253. Thousand Oaks, CA: SAGE, 2018.

Buckley, Chris. "China Enshrines 'Xi Jinping Thought,' Elevating Leader to Mao-like Status." *New York Times*, October 24, 2017. www.nytimes.com.

Callison, William. "Movements of Counter-Speculation: A Conversation with Michel Feher." *Los Angeles Review of Books*, July 12, 2019. https://lareviewofbooks.org.

Cao, Ang, and Pengxiang Li. "We are not Machines: The Identity Construction of Chinese Female Migrant Workers in Online Chat Groups." *Chinese Journal of Communication* 11, no. 3 (2018): 289–305.

Cao, Jin. "Chuanbo Jishu yu Shehui Xingbie: Yi Liu Yi Shanghai de Jiazheng Zhongdian Nügong de Shouji Shiyong Fenxi Weili" [Communication Technology and Gender: An Analysis of Cell Phone use among Migrant Female Domestic Part-Timers in Shanghai]. *Journalism & Communication* 16, no. 1 (2009): 71–77, 109.

Carrico, Kevin. "Producing Purity: An Ethnographic Study of a Neotraditionalist Ladies' Academy in Contemporary China." In *Cultural Politics of Gender and Sexuality in Contemporary China*, edited by Tiantian Zheng, 41–56. Honolulu: University of Hawaii Press, 2016.

Chan, Lik Sam. "Multiple Uses and Anti-Purposefulness on Momo, a Chinese Dating/Social App." *Information, Communication & Society* 23, no. 10 (2020): 1515–1530.

———. *The Politics of Dating Apps: Gender, Sexuality, and Emergent Publics in Urban China*. Cambridge, MA: MIT Press, 2021.

Chang, Che, and Jeremy Goldkorn. "China's 'Big Tech Crackdown': A Guide." The China Project. August 2, 2021. https://thechinaproject.com.

Chen, Jiying. "Jujiao 'Taobaocun': Xiang Xianyu Manyan Cheng Xinxing Chengzhenhua Yangban" [Focus on "Taobao Villages": Spreading to Counties and Becoming a Model of New Urbanization]. *China News Weekly*, September 4, 2013. www.chinanews.com.

Chen, Julie Yujie, and Jack Linchuan Qiu. "Digital Utility: Datafication, Regulation, Labor, and DiDi's Platformization of Urban Transport in China." *Chinese Journal of Communication* 12, no. 3 (2019): 274–289.

Chen, Yujie, Zhifei Mao, and Jack Linchuan Qiu. *Super-Sticky WeChat and Chinese Society*. Bingley, UK: Emerald Group Publishing, 2018.

Chen, Zhaodi, and Dali L. Yang. "Governing Generation Z in China: Bilibili, Bidirectional Mediation, and Online Community Governance." *The Information Society* 39, no. 1 (2023): 1–16.

Cheung, Anne S. Y., and Yongxi Chen. "From Datafication to Data State: Making Sense of China's Social Credit System and its Implications." *Law & Social Inquiry* 47, no. 4 (2022): 1137–1171.

Chib, Arul, Shelly Malik, Rajiv George Aricat, and Siti Zubeidah Kadir. "Migrant Mothering and Mobile Phones: Negotiations of Transnational Identity." *Mobile Media & Communication* 2, no. 1 (2014): 73–93.

Chin, Josh, and Liza Lin. *Surveillance State: Inside China's Quest to Launch a New Era of Social Control*. New York: St. Martin's Press, 2022.

"China Bullet Train Crash 'Caused by Design Flaws.'" *BBC News*, December 28, 2011. www.bbc.com.

China Internet Network Information Center (CNNIC). "The 28th Statistical Report on Internet Development in China." July 2011. www.cnnic.com.cn.

———. "The 32nd Statistical Report on Internet Development in China," July 2013. www.cnnic.com.cn.

———. "The 36th Statistical Report on Internet Development in China," July 2015. www.cnnic.com.cn.

———. "Research Report on Rural Internet Development in 2015." August 2016. www.cnnic.com.cn.

———. "The 44th Statistical Report on Internet Development in China," August 2019. www.cnnic.com.cn.

———. "The 46th Statistical Report on Internet Development in China." September 2020. www.cnnic.com.cn.

———. "The 50th Statistical Report on Internet Development in China." August 2022. www.cnnic.com.cn.

The China Team. "Red Vs are after China's Queer Community." *Protocol*. July 13, 2021. Site no longer active. Summary of article available at https://thechinaproject.com.

"China's Largest Online Domestic Helper Agency Says Labour Shortage a Worry amid Population Decline." *South China Morning Post*, June 22, 2023. https://finance.yahoo.com.

Chitwood, Matthew. "For China, Ending Poverty is Just the Beginning." *Foreign Affairs*, November 19, 2020. www.foreignaffairs.com.

Chong, Gladys Pak Lei. "Who Wants 9-to-5 Jobs? Precarity, (In)security, and Chinese Youths in Beijing and Hong Kong." *The Information Society* 36, no. 5 (2020): 266–278.

Chow, Yiu Fai. "Hong Kong Creative Workers in Mainland China: The Aspirational, the Precarious, and the Ethical." *China Information* 31, no. 1 (2017): 43–62.

———. *Caring in Times of Precarity: A Study of Single Women Doing Creative Work in Shanghai*. Cham, Switzerland: Palgrave Macmillan, 2019.

Chua, Trudy Hui Hui, and Leanne Chang. "Follow me and Like my Beautiful Selfies: Singapore Teenage Girls' Engagement in Self-Presentation and Peer Comparison on Social Media." *Computers in Human Behavior* 55 (2016): 190–197.

Chumley, Lily. *Creativity Class: Art School and Culture Work in Postsocialist China.* Princeton, NJ: Princeton University Press, 2016.

Ci, Jiwei. *Moral China in the Age of Reform.* New York: Cambridge University Press, 2014.

Clark-Parsons, Rosemary. *Networked Feminism: How Digital Media Makers Transformed Gender Justice Movements.* Berkeley: University of California Press, 2022.

Cohen, David. "Making Sense of China's Big Tech Crackdown." Technode. June 28, 2021. https://technode.com.

Coleman, Rebecca. "Bodies, Ethics and Immanent Research: Deleuze's Concept of Affect as Methodology." *Feminist Media Studies* 8, no. 1 (2008): 91–95.

"Connecting the Countryside via E-Commerce: Evidence from China." *American Economic Review: Insights* 3, no. 1 (2021): 35–50.

Conor, Bridget, Rosalind Gill, and Stephanie Taylor. "Gender and Creative Labour." *The Sociological Review* 63 (2015): 1–22.

Coté, Mark, and Jennifer Pybus. "Learning to Immaterial Labour 2.0: MySpace and Social Networks." *ephemera* 7, no. 1 (2007): 88–106.

Couldry, Nick. *Why Voice Matters: Culture and Politics after Neoliberalism.* Thousand Oaks, CA: Sage, 2010.

Couture, Victor, Benjamin Faber, Yizhen Gu, and Lizhi Liu.

Credit Suisse Research Institute. *Global Wealth Databook.* 2021.

Creemers, Rogier. "Cyber China: Upgrading Propaganda, Public Opinion Work and Social Management for the Twenty-first Century." *Journal of Contemporary China* 26, no. 103 (2017): 85–100.

———. "China's Social Credit System: An Evolving Practice of Control." SSRN. May 22, 2018. papers.ssrn.com.

———. "Cybersecurity Law and Regulation in China: Securing the Smart State." *China Law and Society Review* 6, no. 2 (2021): 111–145.

Creemers, Rogier, and Paul Trilio. "Analyzing China's 2021–2025 Informatization Plan: A DigiChina Forum." DigiChina. January 24, 2022. https://digichina.stanford.edu.

Croll, Elisabeth. "New Peasant Family Forms in Rural China." *The Journal of Peasant Studies* 14, no. 4 (1987): 469–499.

———. *Changing Identities of Chinese Women: Rhetoric, Experience and Self-Perception in Twentieth-Century China.* London: Zed Books, 1995.

Curran, James, and Myung-Jin Park, eds. *De-Westernizing Media Studies.* New York: Routledge, 2000.

Cvetkovich, Ann. *An Archive of Feelings: Trauma, Sexuality, and Lesbian Public Cultures.* Durham, NC: Duke University Press, 2003.

Das, Veena. "Engaging the Life of the Other: Love and Everyday Life." In *Ordinary Ethics: Anthropology, Language, and Action,* edited by Michael Lambek, 376–399. New York: Fordham University Press, 2010.

———. "Ordinary Ethics." In *A Companion to Moral Anthropology,* edited by Didier Fassen, 133–149. Malden, MA: Wiley Blackwell, 2012.

de Kloet, Jeroen, Yiu Fai Chow, and Lena Scheen. *Boredom, Shanzhai, and Digitisation in the Time of Creative China*. Amsterdam: Amsterdam University Press, 2019.

de Kloet, Jeroen, Jian Lin, and Jueling Hu. "The Politics of Emotion during COVID-19: Turning Fear into Pride in China's WeChat Discourse." *China Information* 35, no. 3 (2021): 366–392.

de Kloet, Jeroen, Thomas Poell, Guohua Zeng, and Yiu Fai Chow. "The Platformization of Chinese Society: Infrastructure, Governance, and Practice." *Chinese Journal of Communication* 12, no. 3 (2019): 249–256.

Dean, Mitchell. *Governmentality: Power and Rule in Modern Society*. 2nd ed. Thousand Oaks, CA: Sage, 2010.

Deleuze, Gilles. *Spinoza: Practical Philosophy*. Translated by Robert Hurley. San Francisco: City Lights, 1988.

Deleuze, Gilles, and Félix Guattari. *A Thousand Plateaus: Capitalism and Schizophrenia*. Translated by Brian Massumi. Minneapolis: University of Minnesota Press, 1987.

DeLisle, Jacques, Avery Goldstein, and Guobin Yang, eds. *The Internet, Social Media, and a Changing China*. Philadelphia: University of Pennsylvania Press, 2016.

Ding, Rui. "In Shanghai, a New Major Helps Domestic Workers Battle Stigma." Sixth Tone. August 28, 2023. www.sixthtone.com.

Donnellon-May, Genevieve. "China's Push to Advance Rural Revitalization." The Diplomat. February 12, 2022. https://thediplomat.com.

Dosekun, Simidele. "For western girls only? Post-feminism as Transnational Culture." *Feminist Media Studies* 15, no. 6 (2015): 960–975.

Döveling, Katrin, Anu A. Harju, and Denise Sommer. "From Mediatized Emotion to Digital Affect Cultures: New Technologies and Global Flows of Emotion." *Social Media+ Society* 4, no. 1 (2018). https://doi.org/10.1177/2056305117743141.

Du, Fangqin, and Yiping Cai. "A History of the Patriarchal System and Gender Relations." In *Women's Studies in China: Mapping the Social, Economic and Policy Changes in Chinese Women's Lives*, edited by Fangqin Du and Xinrong Zheng, 33–52. Seoul, South Korea: Ewha Women's University Press, 2005.

Duffy, Brooke Erin. *(Not) Getting Paid to Do what you Love*. New Haven, CT: Yale University Press, 2017.

Duggan, Lisa. *The Twilight of Equality? Neoliberalism, Cultural Politics, and the Attack on Democracy*. Boston, MA: Beacon Press, 2003.

Economy, Elizabeth C. *The Third Revolution: Xi Jinping and the New Chinese State*. New York: Oxford University Press, 2018.

Edwards, Louise. "Policing the Modern Woman in Republican China." *Modern China* 26, no. 2 (2000): 115–47.

Ellerman, Mei-Ling. "The Power of Everyday Subordination: Exploring the Silencing and Disempowerment of Chinese Migrant Domestic Workers." *Critical Asian Studies* 49, no. 2 (2017): 187–206.

Erdenebileg, Zolzaya, and Weining Hu. "Made in China 2025: Implications for Foreign Businesses." China Briefing. May 18, 2017. www.china-briefing.com.

Erwin, Kathleen. "Heart-to-Heart, Phone-to-Phone: Family Values, Sexuality, and the Politics of Shanghai's Advice Hotlines." In *The Consumer Revolution in Urban China*, edited by Deborah Davis, 145–170. Berkeley: University of California Press, 2000.

Evans, Harriet. *Women and Sexuality in China: Female Sexuality and Gender since 1949*. New York: Continuum, 1997.

———. "Sexed Bodies, Sexualized Identities, and the Limits of Gender." *China Information* 22, no. 2 (2008): 361–86.

Fan, C. Cindy. "Out to the City and Back to the Village: The Experiences and Contributions of Rural Women Migrating from Sichuan and Anhui." In *On the Move: Women and Rural-to-Urban Migration in Contemporary China*, edited by Arianne M. Gaetano and Tamara Jacka, 177–206. New York: Columbia University Press, 2004.

———. "China's Eleventh Five-Year Plan (2006–2010): From 'Getting Rich First' to 'Common Prosperity.'" *Eurasian Geography and Economics* 47, no. 6 (2006): 708–723.

———. *China on the Move: Migration, the State, and the Household*. New York: Routledge, 2008.

Fan, Jie, Guohua Xiao, and Ling Li. "Zhengyangxian Qianyu Nongcun Funü Liyong Hulianwang Zhifu" [Women in Zhengyang County do Online Business]. *Zhumadian Daily*, January 18, 2007.

Faniyi, Ololade. "Intersectionality in/through Nigeria's Feminist Hashtag Activism." *Communication, Culture & Critique* 16, no. 2 (2023): 110–112.

Feher, Michel. *Rated Agency: Investee Politics in a Speculative Age*. New York: Zone Books, 2018.

Fei, Xiaotong. *From the Soil: The Foundations of Chinese Society*. Translated by Gary G. Hamilton and Zheng Wang. Berkeley: University of California Press, 1992.

Fincher, Leta Hong. *Betraying Big Brother: The Feminist Awakening in China*. London: Verso, 2018.

———. *Leftover Women: The Resurgence of Gender Inequality in China*. 10th anniversary edition. New York: Verso, 2023.

Florida, Richard. *The Rise of the Creative Class*. New York: Basic Books, 2002.

———. *The New Urban Crisis: How our Cities are Increasing Inequality, Deepening Segregation, and Failing the Middle Class—and What We Can Do About It*. New York: Basic Books, 2017.

Fong, Vanessa L. "China's One-Child Policy and the Empowerment of Urban Daughters." *American Anthropologist* 104, no. 4 (2002): 1098–1109.

Fortunati, Leopoldina. "Mobile Telephone and the Presentation of Self." In *Mobile Communications: Re-Negotiation of the Social Sphere*, edited by Rich Ling and Per E. Pedersen, 203–218. London: Springer, 2005.

Foucault, Michel. "Technologies of the Self." In *Technologies of the Self: A Seminar with Michel Foucault*, edited by Luther H. Martin, Huck Gutman, and Patrick H. Hutton, 16–49. Amherst, MA: University of Massachusetts Press, 1988.

———. *The Use of Pleasure: The History of Sexuality*. Vol. 2. Translated by Robert Hurley. New York: Vintage Books, 1990.

———. "Governmentality." In *The Foucault Effect: Studies in Governmentality*, edited by Graham Burchell, Colin Gordon, and Peter Miller, 87–104. Chicago: University of Chicago Press, 1991.

———. *Ethics: Subjectivity and Truth*. Vol. 1 of *Essential Works of Foucault, 1954–1984*. Edited by Paul Rabinow. Translated by Robert Hurley and others. New York: The New Press, 1997.

———. "The Subject and Power." In *Power*. Vol. 3 of *Essential Works of Foucault, 1954–1984*, edited by James Faubion, 326–48. New York: The New Press, 2000.

Fu, Beimeng. "New Film Turns the Tables on China's Infamous 'Chengguan.'" Sixth Tone. April 29, 2021. https://sixthtone.com.

Fu, Diana, and Greg Distelhorst. "Grassroots Participation and Repression under Hu Jintao and Xi Jinping." *The China Journal* 79, no. 1 (2018): 100–122.

Fu, Huiyan, Yihui Su, and Anni Ni. "Selling Motherhood: Gendered Emotional Labor, Citizenly Discounting and Alienation among China's Migrant Domestic Workers." *Gender & Society* 32, no. 6 (2018): 814–36.

Fung, Anthony. "Redefining Creative Labor: East Asian Comparisons." In *Precarious Creativity: Global Media, Local Labor*, edited by Michael Curtin and Kevin Sanson, 200–214. Berkeley: University of California Press, 2016.

Fung, Anthony Y. H., and John Nguyet Erni. "Cultural Clusters and Cultural Industries in China." *Inter-Asia Cultural Studies* 14, no. 4 (2013): 644–656.

Gaetano, Arianne M. "Filial Daughters, Modern Women: Migrant Domestic Workers in Post-Mao Beijing." In *On the Move: Women and Rural-to-Urban Migration in Contemporary China*, edited by Arianne M. Gaetano and Tamara Jacka, 41–79. New York: Columbia University Press, 2004.

———. *Out to Work: Migration, Gender, and the Changing Lives of Rural Women in Contemporary China*. Honolulu: University of Hawaii Press, 2015.

Gan, Nectar. "Chinese Feminists are being Silenced by Nationalist Trolls. Some are Fighting Back." CNN. April 19, 2021. www.cnn.com.

Gao, Ge, and Xiaosui Xiao. "Intercultural/Interpersonal Communication Research in China: A Preliminary Review." In *Chinese Communication Theory and Research: Reflections, New Frontiers, and New Directions*, edited by Wenshan Jia, Xing Lu, and D. Ray Heisey, 21–35. Westport, CT: Ablex, 2002.

Gao, Jia, and Yuanyuan Su. *Social Mobilisation in Post-Industrial China: The Case of Rural Urbanisation*. Northampton, MA: Edward Elgar Publishing, 2019.

Gao, Xiaoxian. "Dangdai Zhongguo Nongcun Laodongli Zhuanyi ji Nongye Nühua Qushi" [The Transfer of Rural Labor Force and the Trend of Feminization of Agriculture in Contemporary China]. *Sociological Studies* 2 (1994): 83–90.

Garde-Hansen, Joanne, and Kristyn Gorton. *Emotion Online: Theorizing Affect on the Internet*. London: Palgrave Macmillan, 2013.

Garrick, John, and Yan Chang Bennett. "'Xi Jinping Thought': Realization of the Chinese Dream of National Rejuvenation?" *China Perspectives* no. 1–2 (2018): 99–105.

Gill, Rosalind. "Postfeminist Media Culture: Elements of a Sensibility." *European Journal of Cultural Studies* 10, no. 2 (2007): 147–166.

Gill, Rosalind, and Andy Pratt. "Precarity and Cultural Work in the Social Factory? Immaterial Labour, Precariousness and Cultural Work." *Theory, Culture & Society* 25, no. 7–8 (2008): 1–30.

Ging, Debbie, and Eugenia Siapera, eds. *Gender Hate Online: Understanding the New Anti-Feminism.* London: Palgrave Macmillan, 2019.

Gleeson, Jessamy. "'(Not) Working 9–5': The Consequences of Contemporary Australian-Based Online Feminist Campaigns as Digital Labour." *Media International Australia* 161, no. 1 (2016): 77–85.

Gleiss, Marielle Stigum. "How Chinese Labour NGOs Legitimize their Identity and Voice." *China Information* 28, no. 3 (2014): 362–381.

Gong, Haomin, and Xin Yang. *Reconfiguring Class, Gender, Ethnicity and Ethics in Chinese Internet Culture.* New York: Routledge, 2017.

Gordon, Colin. "Governmental Rationality: An Introduction." In *The Foucault Effect: Studies in Governmentality*, edited by Graham Burchell, Colin Gordon, and Peter Miller, 1–51. Chicago: University of Chicago Press, 1991.

Gorton, Kristyn. "Theorizing Emotion and Affect: Feminist Engagements." *Feminist Theory* 8, no. 3 (2007): 333–348.

Gould, Deborah. *Moving Politics: Emotion and ACT UP's Fight Against AIDS.* Chicago: University of Chicago Press, 2009.

Gow, Michael. "The Core Socialist Values of the Chinese Dream: Towards a Chinese Integral State." *Critical Asian Studies* 49, no. 1 (2017): 92–116.

Greenhalgh, Susan, and Edwin A. Winckler. *Governing China's Population: From Leninist to Neoliberal Biopolitics.* Stanford, CA: Stanford University Press, 2005.

Gregg, Melissa. *Work's Intimacy.* Malden, MA: Polity, 2011.

Greyser, Naomi. "Beyond the 'Feeling Woman': Feminist Implications of Affect Studies." *Feminist Studies* 38, no. 1 (2012): 84–112.

Grossberg, Lawrence. *We Gotta Get out of this Place: Popular Conservatism and Postmodern Culture.* New York: Routledge, 1992.

Gu, Ming Dong. "Aesthetic Suggestiveness in Chinese Thought: A Symphony of Metaphysics and Aesthetics." *Philosophy East and West* 53, no. 4 (2003): 490–513.

Guo, Shaohua. *The Evolution of the Chinese Internet: Creative Visibility in the Digital Public.* Stanford, CA: Stanford University Press, 2020.

Guo, Yuhua, and Peng Chen. "Digital Divide and Social Cleavage: Case Studies of ICT Usage among Peasants in Contemporary China." *The China Quarterly* 207 (2011): 580–599.

Hafkin, Nancy J., and Sophia Huyer. *Cinderella or Cyberella? Empowering Women in the Knowledge Society.* Bloomfield, CT: Kumarian Press, 2006.

Han, Eileen Le. *Micro-blogging Memories: Weibo and Collective Remembering in Contemporary China.* London: Palgrave Macmillan, 2016.

Han, Hongmei. "Jiazhenggong Weisha gei Guzhu Ding Shouze—Bianyuan Renqun Quanyi Changdao Anli" [Why Domestic Workers Should Have a Code with their

Employers—A Case Study of Public Advocacy for a Marginalized Population]. *China Development Brief* 8 (2013): 58–61.

Han, Ling, and Chengpang Lee. "Nudity, Feminists, and Chinese Online Censorship: A Case Study on the Anti-Domestic Violence Campaign on Sina Weibo." *China Information* 33, no. 3 (2019): 274–293.

Han, Rongbin. *Contesting Cyberspace in China: Online Expression and Authoritarian Resilience.* New York: Columbia University Press, 2018.

Han, Xiao. "Searching for an Online Space for Feminism? The Chinese Feminist Group Gender Watch Women's Voice and its Changing Approaches to Online Misogyny." *Feminist Media Studies* 18, no. 4 (2018): 734–749.

———. "Uncovering the Low-Profile #MeToo Movement: Towards a Discursive Politics of Empowerment on Chinese Social Media." *Global Media & China* 6, no. 3 (2021): 364–380.

Handyside, Sarah, and Jessica Ringrose. "Snapchat Memory and Youth Digital Sexual Cultures: Mediated Temporality, Duration and Affect." *Journal of Gender Studies* 26, no. 3 (2017): 347–360.

Hao, Karen. "Livestreaming Helped China's Farmers Survive the Pandemic. It's Here to Stay." MIT Technology Review. May 6, 2020. www.technologyreview.com.

Hap, Junhui. "Xiaxiang Sinian, Nongcun Taobao 'Quetui' Qianxing" [After Four Years in the Countryside, Rural Taobao is "Limping" Forward]. *IT Times*, September 2, 2018. www.tmtpost.com.

Hardt, Michael. "Affective Labor." *boundary 2* 26, no. 2 (1999): 89–100.

Hardt, Michael, and Antonio Negri. *Empire.* Cambridge, MA: Harvard University Press, 2000.

Harvey, David. *A Brief History of Neoliberalism.* New York: Oxford University Press, 2005.

Harwit, Eric. *China's Telecommunications Revolution.* Oxford: Oxford University Press, 2008.

———. "The Rise and Influence of Weibo (Microblogs) in China." *Asian Survey* 54, no. 6 (2014): 1059–1087.

———. "WeChat: Social and Political Development of China's Dominant Messaging App." *Chinese Journal of Communication* 10, no. 3 (2017): 312–327.

He, Guangye, and Muzhi Zhou. "Gender Difference in Early Occupational Attainment: The Roles of Study Field, Gender Norms, and Gender Attitudes." *Chinese Sociological Review* 50, no. 3 (2018): 339–366.

He, Huaihong. *Social Ethics in a Changing China: Moral Decay or Ethical Awakening?* With a foreword by John L. Thornton and an introduction by Cheng Li. Washington, DC: Brookings Institution Press, 2015.

He, Wei. "Rural Taobao Brings E-Commerce to the Countryside." *China Daily*, December 3, 2017. www.chinadaily.com.cn.

He, Xia. "Li Wuwei Xiangshi Geguo Chuangyi Chanye Fazhan Lishi yu Xianzhuang" [Li Wuwei Explained Various Countries' Creative Industries Historical Development and Present Situation]. *Peoples' Daily Online*, March 15, 2011. www.do-bechina.com.

Hearn, Alison. "Verified: Self-presentation, Identity Management, and Selfhood in the Age of Big Data." *Popular Communication* 15, no. 2 (2017): 62–77.

Hemmings, Claire. "Invoking Affect: Cultural Theory and the Ontological Turn." *Cultural Studies* 19, no. 5 (2005): 548–567.

Herold, David. "Xi Jinping's Internet: Faster, Truer, More Positive and More Chinese." *China: An International Journal* 16, no. 3 (2018): 52–73.

Herold, David Kurt, and Gabriele de Seta. "Through the Looking Glass: Twenty Years of Chinese Internet Research." *The Information Society* 31, no. 1 (2015): 68–82.

Hesmondhalgh, David, and Sarah Baker. *Creative Labour: Media Work in Three Cultural Industries*. New York: Routledge, 2011.

Hillis, Ken, Susanna Paasonen, and Michael Petit, eds. *Networked Affect*. Cambridge, MA: MIT Press, 2015.

Hine, Christine. *Ethnography for the Internet: Embedded, Embodied and Everyday*. New York: Routledge, 2020.

Hird, Derek. "Smile Yourself Happy: *Zheng Nengliang* and the Discursive Construction of Happy Subjects." In *Chinese Discourses on Happiness*, edited by Gerda Wielander and Derek Hird, 106–128. Hong Kong: Hong Kong University Press, 2018.

"His Own Words: The 14 Principles of 'Xi Jinping Thought.'" *BBC Monitoring*, October 24, 2017. https://monitoring.bbc.co.uk.

Hjorth, Larissa, and Sun Sun Lim. "Mobile Intimacy in an Age of Affective Mobile Media." *Feminist Media Studies* 12, no. 4 (2012): 477–484.

Hochschild, Arlie. *The Managed Heart: Commercialization of Human Feeling*. Berkeley: University of California Press, 2003.

Hoffman, Lisa M. *Patriotic Professionalism in Urban China: Fostering Talent*. Philadelphia: Temple University Press, 2010.

Hondagneu-Sotelo, Pierrette. *Doméstica: Immigrant Workers Cleaning and Caring in the Shadows of Affluence*. Berkeley: University of California Press, 2001.

Hong, Yu. *Networking China: The Digital Transformation of the Chinese Economy*. Chicago: University of Illinois Press, 2017.

Hossain, Sarah, and Melanie Beresford. "Paving the Pathway for Women's Empowerment? A Review of Information and Communication Technology Development in Bangladesh." *Contemporary South Asia* 20, no. 4 (2012): 455–469.

Hou, Jiaxi. "Contesting the Vulgar Hanmai Performance from Kuaishou: Online Vigilantism toward Chinese Underclass Youths on Social Media Platforms." In *Introducing Vigilant Audiences*, edited by Daniel Trottier, Rashid Gabdulhakov, and Qian Huang, 49–75. Cambridge, UK: Open Book Publishers, 2020.

Howell, Jude. "Women's Organizations and Civil Society in China: Making a Difference." *International Feminist Journal of Politics* 5, no. 2 (2003): 191–215.

Hu, Yong. *Zhongsheng Xuanhua: Wangluo Shidai de Geren Biaoda yu Gonggong Taolun* [The Great Cacophony: Personal Expression and Public Discussion in the Internet Age]. Nanning: Guangxi Normal University Press, 2008.

Hu, Yong, and Lei Chen. "A Review of Internet-Based Communication Research in China." *Online Media and Global Communication* 1, no. 1 (2022): 124–163.

Hu, Yong, and Qiuxin Chen. "Yuqing: Bentu Gainian yu Bentu Shijian" [Yuqing: A Native Chinese Concept and its Practice]. *Communication & Society* 40 (2017): 33–74.

Hu, Yukun. "Gender and Rural Crises in China's Transition toward a Market Economy." In *Revisiting Gender Inequality: Perspectives from the People's Republic of China*, edited by Qi Wang, Min Dongchao, and Bo Ærenlund Sørensen, 179–202. New York: Palgrave Macmillan, 2016.

Huang, Jingyang, and Kellee S. Tsai. "Securing Authoritarian Capitalism in the Digital Age: The Political Economy of Surveillance in China." *The China Journal* 88, no. 1 (2022): 2–28.

Huang, Ronggui, and Xiaoyi Sun. "Weibo Network, Information Diffusion and Implications for Collective Action in China." *Information, Communication & Society* 17, no. 1 (2014): 86–104.

Huang, Shirlena, and Brenda S. A. Yeoh. "Emotional Labor and Transnational Domestic Work: The Moving Geographies of 'Maid Abuse' in Singapore." *Mobilities* 2, no. 2 (2007): 195–217.

Huang, Xiaoyun, Jessica Vitak, and Yla Tausczik. "'You Don't Have to Know My Past': How WeChat Moments Users Manage Their Evolving Self-Presentation." In *Proceedings of the 2020 CHI Conference on Human Factors in Computing Systems*, 1–13. ACM Digital Library, 2020.

Huang, Yalan. "War on Women: Interlocking Conflicts within *The Vagina Monologues* in China." *Asian Journal of Communication* 26, no. 5 (2016): 466–484.

Huang, Zheping, Jane Zhang, and Sarah Zheng. "What Comes Next as China's Tech Crackdown Winds Down." *The Washington Post*, July 24, 2023. www.washingtonpost.com.

Hwang, Kwang-kuo. "Face and Favor: The Chinese Power Game." *American Journal of Sociology* 92, no. 4 (1987): 944–74.

"The Ideal Chinese Husband: Xi Dada and the Cult of Personality Growing around China's President." *South China Morning Post*, February 29, 2016. www.scmp.com.

iiMedia Life and Travel Research Center. "2022–2023nian Zhongguo Jiazheng Fuwu Hangye Fazhan Pouxi ji Hangye Touzi Jiyu Fenxi Baogao (Baogao Jiexuan)" [2022–2023 China Housekeeping Service Industry Development Analysis and Industry Investment Opportunities Analysis Report (Report Excerpts)]. iiMedia Consulting. July 24, 2022. www.iimedia.cn.

Iqbal, Mansoor. "WeChat Revenue and Usage Statistics." Business of Apps. November 15, 2021. www.businessofapps.com.

Ito, Mizuko. "Introduction." In *Networked Publics*, edited by Kazys Vernelis, 1–14. Cambridge, MA: MIT Press, 2008.

Ito, Mizuko, Daisuke Okabe, and Misa Matsuda. *Personal, Portable, Pedestrian: Mobile Phones in Japanese Life*. Cambridge, MA: MIT Press, 2005.

Jacka, Tamara. *Women's Work in Rural China: Change and Continuity in an Era of Reform*. Cambridge, UK: Cambridge University Press, 1997.

———. *Rural Women in Urban China: Gender, Migration, and Social Change*. Armonk, NY: M. E. Sharpe, 2006.

Jacka, Tamara, and Sally Sargeson. "Introduction: Conceptualizing Women, Gender and Rural Development in China." In *Women, Gender and Rural Development in China*, edited by Tamara Jacka and Sally Sargeson, 1–22. Northampton, MA: Edward Elgar, 2011.

Jackson, Sarah J., Moya Bailey, and Brooke Foucault Welles. *#HashtagActivism: Networks of Race and Gender Justice*. Cambridge, MA: The MIT Press, 2020.

Jeffreys, Elaine, and Gary Sigley. "Governmentality, Governance and China." In *China's Governmentalities: Governing Change, Changing Government*, edited by Elaine Jeffreys, 1–23. London: Routledge, 2009.

Jeffreys, Elaine, and Su Xuezhong. "Governing through Lei Feng: A Mao-era Role Model in Reform-era China." In *New Mentalities of Government in China*, edited by David Bray and Elaine Jeffreys, 30–55. New York: Routledge, 2016.

Jenkins, Henry, with Ravi Purushotma, Margaret Weigal, Katie Clinton, and Alice J. Robison. *Confronting the Challenges of Participatory Culture: Media Education for the 21st Century*. Cambridge, MA: MIT Press, 2009.

Jenkins, Henry, Sam Ford, and Joshua Green. *Spreadable Media: Creating Value and Meaning in a Networked Culture*. New York: NYU Press, 2013.

Ji, Li-Jun, Albert Lee, and Tieyuan Guo. "The Thinking Styles of Chinese People." In *The Oxford Handbook of Chinese Psychology*, edited by Michael Harris Bond, 155–167. New York: Oxford University Press: 2010.

Ji, Siqi. "What is 'Lying Flat,' and Why are Chinese Officials Standing up to it?" *South China Morning Post*, October 24, 2021. www.scmp.com/economy.

Jia, Lianrui, and Tianyang Zhou. "The 'Making' of an Online Celebrity: A Case Study of Chinese Rural Gay Couple An Wei and Wu Yiebin." In *Chinese Social Media: Social, Cultural, and Political Implications*, edited by Mike Kent, Katie Ellis, and Jian Xu, 42–58. New York: Routledge, 2018.

Jiang, Min. "Internet Companies in China: Dancing between the Party Line and the Bottom Line." *Asie.Visions* 47 (January 2012).

"Jiazheng Fuwuye: Xuqiu Wangsheng Haiyao Youzhi Gongji" [Domestic Service Industry: Strong Demand also Needs High Quality Supply]. *Economic Daily*, August 28, 2018. www.xinhuanet.com.

Jin, Yihong. "Mobile Patriarchy: Changes in the Mobile Rural Family." *Social Sciences in China* 32, no. 1 (2011): 26–43.

Judd, Ellen R. *Gender and Power in Rural North China*. Stanford, CA: Stanford University Press, 1994.

Kaiman, Jonathan. "China Arrests 900 in Fake Meat Scandal." *The Guardian*, May 3, 2013. www.theguardian.com.

Kan, Karita, "The New 'Lost Generation': Inequality and Discontent among Chinese Youth." *China Perspectives* no. 2 (2013): 63–73.

Karatzogianni, Athina, and Adi Kuntsman, eds. *Digital Cultures and the Politics of Emotion: Feelings, Affect and Technological Change*. New York: Palgrave Macmillan, 2012.

Keane, Michael. *China's New Creative Clusters: Governance, Human Capital and Investment*. New York: Routledge, 2011a.

———. "Editor's Introduction." In *How Creativity is Changing China*, by Li Wuwei, ii-xix. London: Bloomsbury Academic, 2011b.

———. *Creative Industries in China: Art, Design and Media*. Malden, MA: Polity, 2013.

———. "Unbundling Precarious Creativity in China: 'Knowing-How' and 'Knowing-To.'" In *Precarious Creativity: Global Media, Local Labor*, edited by Michael Curtin and Kevin Sanson, 215–230. Berkeley: University of California Press, 2016.

———. "Creativity, Affordances, and Traditional Chinese Culture." In *Boredom,* Shan-zhai*, and Digitisation in the Time of Creative China*, edited by Jeroen de Kloet, Yiu Fai Chow, and Lena Scheen, 255–273. Amsterdam: Amsterdam University Press, 2019.

Kent, Mike, Katie Ellis, and Jian Xu, eds. *Chinese Social Media: Social, Cultural, and Political Implications*. New York: Routledge, 2018.

Khanna, Tarun, Ryan Allen, Adam Frost, and Wesley Koo. "Rural Taobao: Alibaba's Expansion into Rural E-commerce." Boston: Harvard Business Publishing, 2019. Available online at Case Centre, Case 9-719-433. www.thecasecentre.org.

King, Gary, Jennifer Pan, and Margaret Roberts. "How Censorship in China Allows Government Criticism but Silences Collective Expression." *American Political Science Review* 7, no. 2 (2013): 326–43.

Kipnis, Andrew B. *Producing Guanxi: Sentiment, Self, and Subculture in a North China Village*. Durham, NC: Duke University Press, 1997.

———. "Audit Cultures: Neoliberal Governmentality, Socialist Legacy, or Technologies of Governing?" *American Ethnologist* 35, no. 2 (2008): 275–289.

———. *Governing Educational Desire: Culture, Politics, and Schooling in China*. Chicago: University of Chicago Press, 2011.

Kleinman, Arthur. "The Art of Medicine: Remaking the Moral Person in China: Implications for Health." *Lancet* 375, no. 9720 (2010): 1074–1075.

Kleinman, Arthur, and Joan Kleinman. "Suffering and its Professional Transformation: Toward an Ethnography of Interpersonal Experience." *Culture, Medicine and Psychiatry* 15, no. 3 (1991): 275–301.

Kleinman, Arthur, Yunxiang Yan, Jing Jun, Sing Lee, Everett Zhang, Pan Tianshu, Wu Fei, and Jinhua Guo. "Introduction: Remaking the Moral Person in a New China." In *Deep China: The Moral Life of the Person*, by Arthur Kleinman, Yunxiang Yan, Jing Jun, Sing Lee, Everett Zhang, Pan Tianshu, Wu Fei, and Jinhua Guo, 1–35. Berkeley: University of California Press, 2011.

Knudsen, Britta Timm, and Carsten Stage. "Introduction: Affective Methodologies." In *Affective Methodologies: Developing Cultural Research Strategies for the Study of Affect*, edited by Britta Timm Knudsen and Carsten Stage, 1–22. London: Palgrave Macmillan, 2015.

Koetse, Manya. "Hangzhou Nanny Sets House on Fire, Killing a Mother and her Three Children." What's on Weibo. June 27, 2017. www.whatsonweibo.com.

Koivunen, Anu. "An Affective Turn? Reimaging the Subject of Feminist Theory." In *Working with Affect in Feminist Readings: Disturbing Differences*, edited by Marianne Liljeström and Susanna Paasonen, 8–28. New York: Routledge, 2010.

Kong, Shuyu. "Melodrama for Change: Gender, *Kuqing* and the Affective Articulation of Chinese TV Drama." In *The Political Economy of Affect and Emotion in East Asia*, edited by Jie Yang, 116–133. New York: Routledge, 2014a.

———. *Popular Media, Social Emotion and Public Discourse in Contemporary China*. Routledge, 2014b.

Köppel-Yang, Martina. *Semiotic Warfare: The Chinese Avant-Garde, 1979–1989, A Semiotic Analysis*. Hong Kong: Timezone 8 Limited, 2003.

Kuan, Teresa. *Love's Uncertainty: The Politics and Ethics of Child Rearing in Contemporary China*. Berkeley: University of California Press, 2015.

Kuhn, Robert Lawrence. "Xi Jinping's Chinese Dream." *New York Times*, June 4, 2013. www.nytimes.com.

Kuntsman, Adi. "Introduction: Affective Fabrics of Digital Cultures." In *Digital Cultures and the Politics of Emotion: Feelings, Affect and Technological Change*, edited by Athina Karatzogianni and Adi Kuntsman, 1–17. New York: Palgrave Macmillan, 2012.

Kuo, Lily. "Plight of Chinese Mother of Eight Chained Outside in Winter Causes Public Outrage despite Official Explanations." *The Washington Post*, February 9, 2020. www.washingtonpost.com.

Kuo, Rachel. "Animating Feminist Anger: Economies of Race and Gender in Reaction GIFs." In *Gender Hate Online: Understanding the New Anti-Feminism*, edited by Debbie Ging and Eugenia Siapera, 173–193. Cham, Switzerland: Palgrave Macmillan, 2019.

Kushnir, Khrystyna, Melina Laura Mirmulstein, and Rita Ramalho. "Micro, Small, and Medium Enterprises around the World: How Many are There, and What Affects the Count?" In *MSME Country Indicators Analysis Note*, 1–9. Washington, DC: World Bank Group, 2010.

Laidlaw, James. "Fault Lines in the Anthropology of Ethics." In *Moral Engines: Exploring the Ethical Drives in Human Life*, edited by Cheryl Mattingly, Rasmus Dyring, Maria Louw, and Thomas Schwarz Wentzer, 174–193. New York: Berghan, 2017.

Lam, Oiwan. "China Anti Three-Vulgarity Campaign." Global Voices. August 10, 2010. https://advox.globalvoices.org.

Lambek, Michael. "Introduction." In *Ordinary Ethics: Anthropology, Language, and Action*, edited by Michael Lambek, 1–36. New York: Fordham University Press, 2010.

———. "On the Immanence of the Ethical: A Response to Michael Lempert, 'No Ordinary Ethics.'" *Anthropological Theory* 15, no. 2 (2015a): 128–132.

———. *The Ethical Condition: Essays on Action, Person, and Value*. Chicago: University of Chicago Press, 2015b.

———. "On the Immanence of Ethics." In *Moral Engines: Exploring the Ethical Drives in Human Life*, edited by Cheryl Mattingly, Rasmus Dyring, Maria Louw, and Thomas Schwarz Wentzer, 137–154. New York: Berghan, 2017.

Larson, Christina. "China is Obsessed with Livestreaming and the Censors are Racing to Keep up." *Bloomberg Businessweek*, September 6, 2017. www.bloomberg.com.

Lasén, Amparo. "Affective Technologies—Emotions and Mobile Phones." *Receiver* 11 (2004).

Lee, Melanie. "Fake Apple Store in China even Fools Staff." Reuters. July 21, 2011. www.reuters.com.

Lempert, Michael. "No Ordinary Ethics." *Anthropological Theory* 13, no. 4 (2013): 370–393.

———. "Ethics without Immanence: A Reply to Michael Lambek." *Anthropological Theory* 15, no. 2 (2015): 133–140.

Lessig, Lawrence. *Remix: Making Art and Commerce Thrive in the Hybrid Economy*. New York: Penguin, 2008.

Li, Anthony H. F. "E-commerce and Taobao Villages. A Promise for China's Rural Development?" *China Perspectives* no. 3 (2017): 57–62.

Li, Bin. "*Renmin Ribao* Pinglunyuan Guancha: Rang 'Xinnongmin' Gengyu Xiwang Tianye" [*People's Daily* Commentator Observes: Let "New Farmers" Cultivate the Fields of Hope]. *People's Daily*, November 28, 2017. http://opinion.people.com.cn.

Li, Cheng. "Introduction: Bringing Ethics back into Chinese Discourse." In *Social Ethics in a Changing China: Moral Decay or Ethical Awakening?* by Huaihong He, xv-xxxv. Washington, DC: Brookings Institution Press, 2015.

Li, Guoxiang. "Ruhe Lijie Xiangcun Zhenxing Zhanlüe de Wuge Zhenxing" [How to Understand the "Five Revitalizations" of the Rural Revitalization Strategy]. Banyuetan. November 19, 2018. www.banyuetan.org.

Li, Jun, and Xiaoqin Li. "Media as a Core Political Resource: The Young Feminist Movements in China." *Chinese Journal of Communication* 10, no. 1 (2017): 54–71.

Li, Keqiang. "Chinese Premier Li Keqiang's Speech at Davos." World Economic Forum. January 23, 2015. www.weforum.org.

Li, Miao, Chris K. K. Tan, and Yuting Yang. "*Shehui Ren*: Cultural Production and Rural Youths' Use of the *Kuaishou* Video-sharing App in Eastern China." *Information, Communication & Society* 23, no. 10 (2020): 1499–1514.

Li, Sipan. "Weibo Nüquan de Qianshi Jinsheng: Cong 'Zhengzhi Zhengqu' dao 'Shangye'" [The Past and Present Life of Weibo Feminism: From "Political Correctness" to "Commercial Correctness"]. *The Paper*, June 16, 2020. www.thepaper.cn.

Li, Wuwei. "Guanyu wo Guo Wenhua Chuangychanye Fazhan de Sikao" [On the Development of Culture and Creative Industries in China]. people.com.cn. November 23, 2009. http://theory.people.com.cn.

———. *How Creativity is Changing China*. Translated by Michael Keane. London: Bloomsbury Academic, 2011.

Li, Xiaojiang. "With What Discourse do we Reflect on Chinese Women? Thoughts on Transnational Feminism in China." In *Spaces of Their Own: Women's Public Sphere in Transnational China*, edited by Mayfair Mei-hua Yang, 266–271. Minneapolis: University of Minnesota Press, 1999.

Li, Yan, and Jie Bai. "Xinshengdai Nüxing Nongmingong yu Weibo Chuanbo—Yi Sige Butong Dingwei de Weibo Weili" [New Generation Women Migrant Workers and their Communication through Microblog]. *Journal of Yangzhou University (Humanities and Social Sciences)* 19, no. 6 (2015): 69–77.

Li, Ying. "Bloody Brides in Abuse Protest." *Global Times*, February 14, 2012. www.globaltimes.cn.

Li, Yunkun, and Junhui Hao. "IT Shibao Diaocha: Ali Nongcun Taobao Shengji Zhan-lüe" [IT Times: In-depth Investigation of Alibaba's Rural Taobao Upgrade Strategy]. Tech Sina. December, 2018. https://tech.sina.com.cn.

Li, Zehou. *The Chinese Aesthetic Tradition*. Translated by Maija Bell Samei. Honolulu: University of Hawaii Press, 2009.

Liang, Chenyu. "Renren, Once China's Answer to Facebook, is a Digital Ghost Town." Sixth Tone. September 3, 2018. www.sixthtone.com.

Liang, Fan, and Yuchen Chen. "The Making of 'Good' Citizens: China's Social Credit Systems and Infrastructures of Social Quantification." *Policy & Internet* 14, no. 1 (2022): 114–135.

Liang, Fan, Yuchen Chen, and Fangwei Zhao. "The Platformization of Propaganda: How Xuexi Qiangguo Expands Persuasion and Assesses Citizens in China." *International Journal of Communication* 15 (2021): 1855–1874.

Liao, Rita. "China Hits Roadblock as it Spends Billions to Help Farmers Sell Online." TechinAsia. March 2018. www.techinasia.com.

Liao, Sara. "The Platformization of Misogyny: Popular Media, Gender Politics, and Misogyny in China's State-Market Nexus." Media, Culture & Society 46, no. 1 (2024): 191–203.

Lim, Louisa. *The People's Republic of Amnesia: Tiananmen Revisited*. New York: Oxford University Press, 2014.

Lin, Geng, Xiaoru Xie, and Zuyi Lü. "Taobao Practices, Everyday Life and Emerging Hybrid Rurality in Contemporary China." *Journal of Rural Studies* 47 (2016): 514–523.

Lin, Jian. *Chinese Creator Economies: Labor and Bilateral Creative Workers*. New York: NYU Press, 2023.

Lin, Jian, and Jeroen de Kloet. "Platformization of the Unlikely Creative Class: *Kuaishou* and Chinese Digital Cultural Production." *Social Media + Society* 5, no. 4 (2019). https://doi.org/10.1177/2056305119883430.

Lindtner, Silvia M. *Prototype Nation: China and the Contested Promise of Innovation*. Princeton, NJ: Princeton University Press, 2020.

Liu, Bohong. "The All-China Women's Federation and Women NGOs." In *Chinese Women Organizing: Cadres, Feminists, Muslims, Queers*, edited by Ping-Chun Hsiung, Maria Jaschok, and Cecelia Milwertz, with Red Chan, 141–157. Oxford: Berg, 2001.

Liu, Chiu-Wan. "Return Migration, Online Entrepreneurship and Gender Performance in the Chinese 'Taobao families.'" *Asia Pacific Viewpoint* 61, no. 3 (2020): 478–493.

Liu, Chun. "The Myth of Informatization in Rural Areas: The Case of China's Sichuan Province." *Government Information Quarterly* 29, no. 1 (2012): 85–97.

Liu, Fengshu. *Urban Youth in China: Modernity, the Internet and the Self.* New York: Routledge, 2011.

Liu, Juan, and Jingzhong Ye. "Nongcun Hulianwang de Yongyou he Shiyong: Youguan Fazhan de Sikao" [Ownership and Use of the Internet in Rural Areas: Reflections on Development]. *Journal of China Agricultural University: Social Sciences Edition* 27, no. 4 (2010): 70–78.

Liu, Xiaoxuan, and Tian Tan. "Weibo Chang li Nongmingong Quanyi Huayu de Jiushu You duo Yuan? Yi Xinlang Weibo Weili" [To What Extent can Microblogs Grant Migrant Workers the Right to Express Themselves? A Case Study of Sina Weibo]. *Journal of Guangdong University of Foreign Studies* 23, no. 6 (2007): 55–59.

Liu, Xin. *In One's Own Shadow: An Ethnographic Account of the Condition of Post-reform Rural China.* Berkeley: University of California Press, 2000.

Liu, Yujing, and Louise Moon. "Sexist Adverts and Fear of Maternity Leave: How Gender Inequality in China's Job Market is 'Getting Worse.'" *South China Morning Post*, April 23, 2018. www.scmp.com.

Looney, Kristen E. "China's Campaign to Build a New Socialist Countryside: Village Modernization, Peasant Councils, and the Ganzhou Model of Rural Development." *The China Quarterly* 224 (2015): 909–932.

Lorde, Audre. *Sister Outsider: Essays and Speeches.* Berkeley: Crossing Press, 2007.

Low, Zoe. "Founder of US$14.5 billion Chinese Pyramid Scheme Shanxinhui Jailed for 17 Years." *South China Morning Post*, December 14, 2018. www.scmp.com.

Luo, Wei. "Aching for the Altered Body: Beauty Economy and Chinese Women's Consumption of Cosmetic Surgery." *Women's Studies International Forum* 38 (2013): 1–10.

Luthra, Rashmi. "Recovering Women's Voice: Communicative Empowerment of Women of the South." In *Communication Yearbook 27*, edited by Pamela Kalbfleisch, 45–65. Mahwah, NJ: Lawrence Erlbaum Associates, 2003.

Lutz, Catherine A., and Lila Abu-Lughod, eds. *Language and the Politics of Emotion.* Cambridge: Cambridge University Press, 1990.

Ma, Dan. "Beijingshi Jiazhenggong de Jiuye Tezheng: 'Fei Zhenggui Jiuye' yu 'Qinggan Laodong'" [Employment Characteristics of Domestic Workers in Beijing: "Informal Employment" and "Emotional Labor"]. *China Science and Technology Information* 6 (2010): 30.

Ma, Lin, and Hongyan Li. "Xiangxun Hulianwang Fazhan Yanjiu Zhuangkuang Gaishu" [An Overview of Research on ICTs for Rural Development]. *Journalism* 6, (2011): 78–81.

MacDonald, Tom. *Social Media in Rural China.* London: UCL Press, 2016.

Madianou, Mirca. "Migration and the Accentuated Ambivalence of Motherhood: The Role of ICTs in Filipino Transnational Families." *Global Networks* 12, no. 3 (2012): 277–295.

Marolt, Peter, and David Kurt Herold, eds. *China Online: Locating Society in Online Spaces.* New York: Routledge, 2015.

Massumi, Brian. *Parables for the Virtual*. Durham, NC: Duke University Press, 2002.

Mattingly, Cheryl, and Jason Throop. "The Anthropology of Ethics and Morality." *Annual Review of Anthropology* 47 (2018): 475–492.

McDonald, Joe, and Zen Soo. "Protesting Workers Beaten at Chinese iPhone Factory." Associated Press. November 23, 2022. https://apnews.com.

McDougall, Bonnie S. *Mao Zedong's "Talks at the Yan'an Conference on Literature and Art": A Translation of the 1943 Text with Commentary*. Ann Arbor: University of Michigan Press, 1980/2020.

McRobbie, Angela. "Post-Feminism and Popular Culture." *Feminist Media Studies* 4, no. 3 (2004): 255–264.

———. *Be Creative: Making a Living in the New Culture Industries*. Malden, MA: Polity, 2018.

Mendes, Kaitlynn, Jessica Ringrose, and Jessalynn Keller. *Digital Feminist Activism: Girls and Women Fight Back against Rape Culture*. New York: Oxford University Press, 2019.

Meng, Bingchun. *The Politics of Chinese Media: Consensus and Contestation*. New York: Palgrave Macmillan, 2018.

Meng, Bingchun, and Yanning Huang. "Patriarchal Capitalism with Chinese Characteristics: Gendered Discourse of 'Double Eleven' Shopping Festival." *Cultural Studies* 31, no. 5 (2017): 659–684.

Miao, Weishan, Min Jiang, and Yunxia Pang. "Historicizing Internet Regulation in China: A Meta-analysis of Chinese Internet Policies (1994–2017)." *International Journal of Communication* 15 (2021): 2003–2026.

Milwertz, Cecilia. *Beijing Women Organizing for Change: A New Wave of the Chinese Women's Movement*. Copenhagen: Nordic Institute of Asian Studies, 2002.

Ministry of Commerce of the People's Republic of China. "Xiliejieduliu: <Dianzi Shangwufa> Cujin Zhongdian Fangxiang: Nongcun Dianshang yu Dianshang Fupin" [Series Interpretation 6: Key Directions Promoted by the E-Commerce Law: Rural E-Commerce and E-Commerce Poverty Alleviation]. January 1, 2019. www.mofcom.gov.cn.

Ministry of Industry and Information Technology of the People's Republic of China. "Guanyu Shishi Kuandai Zhongguo 2013 Zhuanxiang Xingdong de Yijian" [Opinions on the Implementation of the 2013 Special Action for Broadband China]. April 17, 2013. www.gov.cn.

Mo, Rong, Jianfei Zhang, Saier Ding, and Xueqiu Yang. *Zhongguo Jiazheng Fuwu ye Fazhan Baogao* [Annual Report on China's Domestic Service Industry Development]. SSAP. December 12, 2022. www.pishu.com.cn.

Mohapatra, Sandeep, Scott Rozelle, and Rachel Goodhue. "The Rise of Self-Employment in Rural China: Development or Distress?" *World Development* 35, no. 1 (2007): 163–181.

Montgomery, Lucy. *China's Creative Industries: Copyright, Social Network Markets and the Business of Culture in a Digital Age*. Northampton, MA: Edward Elgar Publishing, 2010.

Morgan, George, and Pariece Nelligan. "Labile Labour—Gender, Flexibility and Creative Work." *The Sociological Review* 63 (2015): 66–83.

Mumporeze, Nadine, and Michael Prieler. "Gender Digital Divide in Rwanda: A Qualitative Analysis of Socioeconomic Factors." *Telematics and Informatics* 34, no. 7 (2017): 1285–1293.

Murphy, Rachel. *How Migrant Labor is Changing Rural China.* Cambridge, UK: Cambridge University Press, 2002.

Myers, Steven Lee. "A Clean-up of 'Holes in the Wall' of China's Capital." *New York Times*, July 17 2017. www.nytimes.com.

Nathan, Andrew J. "Authoritarian Resilience." *Journal of Democracy* 14, no. 1 (2003): 6–17.

National Bureau of Statistics. "Statistical Communiqué of the People's Republic of China on the 2018 National Economic and Social Development." February 28, 2019. www.stats.gov.cn/english.

National Radio and Television Administration. "Guojia Guangbo Dianshi Zongju Bangong Ting Guanyu Jinyibu Jiaqiang Wenyi Jiemu jiqi Renyuan Guanli de Tongzhi" [Notice from the General Office of the National Radio and Television Administration on Further Strengthening the Management of Cultural and Artistic Programs and their Personnel]. September 2, 2021. www.nrta.gov.cn.

Negro, Gianluigi. *The Internet in China: From Infrastructure to a Nascent Civil Society.* Cham, Switzerland: Palgrave Macmillan, 2017.

Negroponte, Nicholas. *Being Digital.* New York: Vintage Books, 1996.

Ngai, Sianne. *Ugly Feelings.* Cambridge, MA: Harvard University Press, 2005.

Ni, Vincent. "The Rise of 'Bai Lan': Why China's Frustrated Youth are Ready to 'Let it Rot.'" *The Guardian*, May 25, 2022a. www.theguardian.com.

———. "China Charges 28 People over Restaurant Attack on Group of Women." *The Guardian.* August 29, 2022b. www.theguardian.com.

Nie, Lei, Cuixiao Fu, and Dan Cheng. "Weixin Pengyouquan: Shehui Wangluo Shijiaoxia de Xunli Shequ, Xinwen Jizhe" [WeChat Moments: A Virtual Community from a Social Network Perspective]. *News Reporter* 5 (2013): 71–75.

Ning, Rundong, and David A. Palmer. "Ethics of the Heart: Moral Breakdown and the Aporia of Chinese Volunteers." *Current Anthropology* 61, no. 4 (2020): 395–417.

Nonini, Donald M. "Is China Becoming Neoliberal?" *Critique of Anthropology* 28, no. 2 (2008): 145–176.

O'Connor, Justin, and Xin Gu. *Red Creative: Culture and Modernity in China.* Bristol, UK: Intellect Books, 2020.

Ong, Aihwa, and Li Zhang. "Introduction: Privatizing China: Powers of the Self, Socialism from Afar." In *Privatizing China: Socialism from Afar*, edited by Li Zhang and Aihwa Ong, 1–19. Ithaca, NY: Cornell University Press, 2008.

Ong, Lynette H. *Outsourcing Repression: Everyday State Power in Contemporary China.* New York: Oxford University Press, 2022.

Oreglia, Elisa. "ICT and (Personal) Development in Rural China." *Information Technologies and International Development* 10, no. 3 (2014): 19–30.

Oreglia, Elisa, and Joseph Kaye. "A Gift from the City: Mobile Phones in Rural China." In *Proceedings of the ACM 2012 Conference on Computer Supported Cooperative Work*, 137–146. New York: Association for Computing Machinery, 2012.

Osburg, John. "Morality and Cynicism in a 'Grey' World." In *Irony, Cynicism and the Chinese State*, edited by Hans Steinmüller and Susanne Brandtstädter, 47–62. New York: Routledge, 2016.

Ouellette, Laurie, and James Hay. *Better Living through Reality TV*. Malden, MA: Blackwell, 2007.

"Outcry over Treatment of Female Staff on Front Line of China's Coronavirus Fight." *The Straits Times*, March 7, 2020. www.straitstimes.com.

Oxfeld, Ellen. *Drink Water, but Remember the Source: Moral Discourse in a Chinese Village*. Berkeley: University of California Press, 2010.

Paetsch, Von Martin. "Van Gogh from the Sweatshop." *Der Spiegel*, August 23, 2006. www.spiegel.de.

"Painting by Numbers." *The Economist*, June 8, 2006. www.economist.com.

Palmer, David A., and Fabian Winiger. "Neo-socialist Governmentality: Managing Freedom in the People's Republic of China." *Economy and Society* 48, no. 4 (2019): 554–578.

Pang, Laikwan. *Creativity and its Discontents: China's Creative Industries and Intellectual Property Rights Offenses*. Durham, NC: Duke University Press, 2012.

Papacharissi, Zizi. *Affective Publics: Sentiment, Technology, and Politics*. New York: Oxford University Press, 2015.

Parreñas, Rhacel Salazar. "Long Distance Intimacy: Class, Gender and Intergenerational Relations between Mothers and Children in Filipino Transnational Families." *Global Networks* 5, no. 4 (2005): 317–336.

———. *The Force of Domesticity: Filipina Migrants and Globalization*. New York: NYU Press, 2008.

Peck, Jamie. "The Creativity Fix." *Eurozine*, June 28, 2007. www.eurozine.com.

Pedwell, Carolyn, and Anne Whitehead. "Affecting Feminism: Questions of Feeling in Feminist Theory." *Feminist Theory* 13, no. 2 (2012): 115–129.

Peng, Altman Yuzhu. "Sharing Food Photographs on Social Media: Performative Xiaozi Lifestyle in Young, Middle-Class Chinese Urbanites' WeChat 'Moments.'" *Social Identities* 25, no. 2 (2019): 269–287.

———. *A Feminist Reading of China's Digital Public Sphere*. Switzerland: Springer International Publishing, 2020.

Peng, Weiying, and Wilfred Yang Wang. "Buying on Weixin/WeChat: Proposing a Sociomaterial Approach of Platform Studies." *Media, Culture & Society* 43, no. 5 (2021): 945-956.

Perry, Elizabeth. "Moving the Masses: Emotion Work in the Chinese Revolution." *Mobilization: An International Quarterly* 7, no. 2 (2002): 111–128.

Petit, Michael. "Digital Disaffect: Teaching through Screens." In *Networked Affect*, edited by Ken Hillis, Susanna Paasonen, and Michael Petit, 169–183. Cambridge, MA: MIT Press, 2015.

Pils, Eva. "From Authoritarian Development to Totalist Urban Reordering: The Daxing Forced Evictions Case." *China Information* 34, no. 2 (2020): 270–290.

Plantin, Jean-Christophe, and Gabriele de Seta. "WeChat as Infrastructure: The Techno-nationalist Shaping of Chinese Digital Platforms." *Chinese Journal of Communication* 12, no. 3 (2019): 257–273.

Platt, Maria, Brenda S. A. Yeoh, Krisel Anne Acedera, Khoo Choon Yen, Grace Baey, and Theodore Lam. "Renegotiating Migration Experiences: Indonesian Domestic Workers in Singapore and Use of Information Communication Technologies." *New Media & Society* 19, no. 10 (2016): 2207–23.

Postill, John, and Sarah Pink. "Social Media Ethnography: The Digital Researcher in a Messy Web." *Media International Australia* 145, no. 1 (2012): 123–134.

Powell, Sian. "Becky Li, Chinese KOL who Sold 100 Cars Online in Four Minutes, on Brands' New Tool to Gauge Influence: 'They are Cruel.'" *South China Morning Post*, February 5, 2021. www.scmp.com.

Probyn, Elspeth. *Blush: Faces of Shame*. Minneapolis: University of Minnesota Press, 2005.

Pun, Ngai. *Made in China: Women Factory Workers in a Global Workplace*. Durham, NC: Duke University Press, 2005.

Pupavac, Vanessa. "Therapeutic Governance: Psycho-Social Intervention and Trauma Risk Management." *Disasters* 25, no. 4 (2001): 358–372.

Qi, Jiaqi, Xiaoyong Zheng, and Hongdong Guo. "The Formation of Taobao Villages in China." *China Economic Review* 53 (2019): 106–127.

Qian, Jun-Yue. "Fei Zhengguixing de 'Zhengguihua': Laodongli Shichang Fenge yu Jiazheng Fuwu Qiye—Jiyu 2019nian Si Chengshi Jiazhenggong Diaocha" ["Formalizing" the Informal: Labor Market Segmentation and Domestic Service Enterprises: Based on the 2019 Survey of Domestic Workers in Four Cities]. *Collection of Women's Studies* 169, no. 1 (2022): 35–51.

Qiang, Zhen-Wei Christine, Asheeta Bhavnani, Nagy K. Hanna, Kaoru Kimura, and Randeep Sudan. *Rural Informatization in China*. Washington, DC: World Bank, 2009.

Qin, Yong, and Yingfeng Fang. "The Effects of E-Commerce on Regional Poverty Reduction: Evidence from China's Rural E-Commerce Demonstration County Program." *China & World Economy* 30, no. 3 (2022): 161–186.

Qiu, Jack Linchuan. *Working-class Network Society: Communication Technology and the Information Have-less in Urban China*. Cambridge, MA: MIT Press, 2009.

———. *Goodbye iSlave: A Manifesto for Digital Abolition*. Urbana: University of Illinois Press, 2016.

Rakow, Lana. and Vija Navarro. "Remote Mothering and the Parallel Shift: Women Meet the Cellular Telephone." *Critical Studies in Media Communication* 10, no. 2 (1993): 144–57.

Rentschler, Carrie A., and Samantha C. Thrift. "Doing Feminism in the Network: Networked Laughter and the 'Binders Full of Women' Meme." *Feminist Theory* 16, no. 3 (2015): 329–359.

Richaud, Lisa. 2021. "Introduction: The Politics of Negative Affects in Post-Reform China." *HAU: Journal of Ethnographic Theory* 11, no. 3 (2021): 901–914.

Roberts, Margaret. *Censored: Distraction and Diversion Inside Chinas Great Firewall*. Princeton, NJ: Princeton University Press, 2018.

Rofel, Lisa. *Desiring China: Experiments in Neoliberalism, Sexuality, and Public Culture in China*. Durham, NC: Duke University Press, 2007.

Rose, Nikolas. *Powers of Freedom: Reframing Political Thought*. New York: Cambridge University Press, 1999.

Rosen, Stanley. "Contemporary Chinese Youth and the State." *Journal of Asian Studies* 68, no. 2 (2009): 359–369.

Rottenberg, Catherine. *The Rise of Neoliberal Feminism*. New York: Oxford University Press, 2018.

Sampson, Tony, Stephen Maddison, and Darren Ellis, eds. *Affect and Social Media: Emotion, Mediation, Anxiety and Contagion*. Lanham, MD: Rowman & Littlefield, 2018.

Savat, David, and Mark Poster, eds. *Deleuze and New Technology*. Edinburgh: Edinburgh University Press, 2009.

Seigworth, Gregory J., and Melissa Gregg. "An Inventory of Shimmers." In *The Affect Theory Reader*, edited by Melissa Gregg and Gregory J. Seigworth, 1–25. Durham, NC: Duke University Press, 2010.

Senft, Theresa M., and Nancy K. Baym. "What Does the Selfie Say? Investigating a Global Phenomenon." *International Journal of Communication* 9, no. 1 (2015): 1588–1606.

Shambaugh, David. *China's Leaders: From Mao to Now*. Medford, MA: Polity, 2021.

Shi, Liang, and Bao Ai. "Zhongguo Zhongbu Nongcun 'Shuzi Honggou' Yanjiu—Yi Anhui Sheng, Hubei Sheng Liang Cunzhuang Weili" [The Digital Divide in the Central Chinese Countryside—A Case Study of Two Villages in Anhui and Hubei]. *Youth Journalist* 26 (2013): 25–26.

Shi, Yuzhi. "Zhongguo Meng Qubie yu Meiguo Meng de Qi Da Tezheng" [Seven Major Differences between the China Dream and the American Dream]. *Renmin Luntan*, May 27, 2013. Archived copy: www.qstheory.cn.

Shih, Shu-mei. "Towards an Ethics of Transnational Encounters, Or 'When' does a 'Chinese' Woman Become a 'Feminist'?" In *Dialogue and Difference: Feminisms Challenge Globalization*, edited by Marguerite Waller and Sylvia Marcos, 3–28. New York: Palgrave Macmillan, 2005.

Shirazi, Farid. "Information and Communication Technology and Women Empowerment in Iran." *Telematics and Informatics* 29, no. 1 (2012): 45–55.

Sinnreich, Aram. *Mashed Up: Music, Technology, and the Rise of Configurable Culture*. Amherst: University of Massachusetts Press, 2010.

Slack, Jennifer Daryl, and J. MacGregor Wise. *Culture + Technology: A Primer.* 2nd ed. New York: Peter Lang, 2015.

Song, Jing. "Women and Self-employment in Post-socialist Rural China: Side Job, Individual Career or Family Venture." *China Quarterly* 221 (2015): 229–242.

Song, Shaopeng. "'Returning Home' or 'Being Returned Home'? The Debate over Women Returning to the Home and Changing Values." In *Revisiting Gender Inequality: Perspectives from the People's Republic of China,* edited by Qi Wang, Min Dongchao, and Bo Ærenlund Sørensen, 59–84. New York: Palgrave MacMillan, 2016.

Spakowski, Nicola. "'Gender' Trouble: Feminism in China under the Impact of Western Theory and the Spatialization of Identity." *positions: east asia cultures critique* 19, no. 1 (2011): 31–54.

———. "Socialist Feminism in Postsocialist China." *positions: asia critique* 26, no. 4 (2018): 561–592.

Spelman, Elizabeth. "Anger and Insubordination." In *Women, Knowledge, and Reality: Explorations in Feminist Philosophy,* edited by Ann Garry and Marilyn Pearsall, 263–274. Boston: Unwin Hyman, 1989.

Spencer, Richard. "Two Sentenced to Death over China Melamine Milk Scandal." *Telegraph* online, January 22, 2008. www.telegraph.co.uk.

Spinoza, Baruch. *Ethics.* Edited and translated by Edwin Curley. New York: Penguin, 1996.

Stafford, Charles. "Ordinary Ethics in China Today." In *Ordinary Ethics in China,* edited by Charles Stafford, 3–25. New York: Routledge, 2013.

State Council of the People's Republic of China. "Zhongban, Guoban Yinfa '2006–2020 Nian Xinxihua Fazhan Zhanlüe" [The General Office of the CPC Central Committee and the General Office of the State Council Issue "2006–2020 Informatization Development Strategy"]. May 8, 2006. www.gov.cn.

———. "'Made in China 2025' Plan Issued." May 19, 2015a. english.www.gov.cn.

———. "China Boosts Mass Entrepreneurship and Innovation." June 16, 2015b. english. www.gov.cn.

———. "Guideline on Measures to Boost Mass Entrepreneurship and Innovation." June 16, 2015c. english.www.gov.cn.

———. "China Unveils Internet Plus Action Plan to Fuel Growth." July 4, 2015d. english.www.gov.cn.

———. "Full Transcript of the State Council Policy Briefing on Feb. 5, 2016." February 5, 2016. english.www.gov.cn.

Steinmüller, Hans. *Communities of Complicity: Everyday Ethics in Rural China.* New York and London: Berghahn Books, 2013.

Stevenson, Alexandra. "China's Male Leaders Signal to Women That Their Place Is in the Home." *New York Times,* November 3, 2023. www.nytimes.com.

Stewart, Kathleen. *Ordinary Affects.* Durham, NC: Duke University Press, 2007.

———. "Atmospheric Attunements." *Environment and Planning D: Society and Space* 29, no. 3 (2011): 445–453.

Suda, Kimiko. "A Room of One's Own: Highly Educated Migrants' Strategies for Creating a Home in Guangzhou." *Population, Space and Place* 22, no. 2 (2016): 146–157.

Sun, Liping. "2012 Nian Shi Da Liu Xing yu Fabu: 'Zheng nengliang' wei Ju Bangshou" [Top Ten Catchphrases of 2012 Published: "Positive Energy" Tops the List]. *China News*, December 30, 2012. www.chinanews.com.

Sun, Longji. *Zhongguo Wenhua de Shenceng Jiegou* [The Deep Structure of Chinese Culture]. Nanning: Guangxi Normal University Press, 2011.

Sun, Ping, Guoning Zhao, Zhen Liu, Xiaoting Li, and Yunze Zhao. "Toward Discourse Involution within China's Internet: Class, Voice, and Social Media." *New Media & Society* 24, no. 5 (2020): 1033–1052.

Sun, Wanning. *Maid in China: Media, Morality, and the Cultural Politics of Boundaries.* New York: Routledge, 2009.

———. *Subaltern China: Rural Migrants, Media, and Cultural Practices.* Lanham, MD: Rowman & Littlefield, 2014.

Sundararajan, Louise. "The Function of Negative Emotions in the Confucian Tradition." In *The Positive Side of Negative Emotions*, edited by Gerrod W. Parrot, 179–197. New York: Guilford Publications, 2014.

———. *Understanding Emotion in Chinese Culture.* New York: Springer, 2015.

Sundararajan, Louise, and James R. Averill. "Creativity in the Everyday: Culture, Self, and Emotions." In *Everyday Creativity and New Views of Human Nature: Psychological, Social, and Spiritual Perspectives*, edited by Ruth Richards, 195–220. Washington, DC: American Psychological Association, 2007.

Surman, Jan, and Ella Rossman. "New Dissidence in Contemporary Russia: Students, Feminism and New Ethics." *New Perspectives: Interdisciplinary Journal of Central & East European Politics & International Relations* 30, no. 1 (2022): 27–46.

Svensson, Marina. "Voice, Power and Connectivity in China's Microblogosphere: Digital Divides on Sina Weibo." *China Information* 28, no. 2 (2014): 168–188.

Sy, Nicole. "Analysis: Chinese E-commerce Companies Head to the Countryside." *South China Morning Post*, May 14, 2015. www.scmp.com.

Szablewicz, Marcella. *Mapping Digital Game Culture in China: From Internet Addicts to Esports Athletes.* Switzerland: Palgrave Macmillan, 2020.

Tan, Chris K. K., Jie Wang, Shengyuan Wangzhu, Jinjing Xu, and Chunxuan Zhu. "The Real Digital Housewives of China's Kuaishou Video-sharing and Live-streaming App." *Media, Culture & Society* 42, no. 7–8 (2020): 1243–1259.

Tan, Jia. *Digital Masquerade: Feminist Rights and Queer Media in China.* New York: NYU Press, 2023.

Tang, Guangxin. "Nongjia Funü Keji Xingjia" [Rural Women Support Family via Online Business]. *China Women's News*, June 3, 2012, A02.

Tang, Ling. "Burning out in Emotional Capitalism: Appropriation of *Ganqing* and *Renqing* in the Chinese Platform Economy." *Journal of Sociology* 59, no. 2 (2023): 421–436.

Tang, Ningjing, Lei Tao, Bo Wen, and Zhicong Lu. "Dare to Dream, Dare to Livestream: How E-commerce Livestreaming Empowers Chinese Rural Women."

In *Proceedings of the 2022 CHI Conference on Human Factors in Computing Systems*, 1–13. ACM Digital Library, 2022.

Tatlow, Didi Kirsten. "For China, a New Kind of Feminism." *New York Times*, September 17, 2013. www.nytimes.com.

Tefft, Sheila. "Painters Flock to Beijing Colony." *Christian Science Monitor*, August 18, 1993. www.csmonitor.com.

Terranova, Tiziana. "Free Labor: Producing Culture for the Digital Economy." *Social Text* 18, no. 2 (2000): 33–58.

Thomas, Minu, and Sun Sun Lim. "ICT Use and Female Migrant Workers in Singapore." In *Mobile Communication: Dimensions of Social Policy*, edited by James Katz, 175–190. New Brunswick, NJ: Transaction, 2011.

Thornton, Sarah. *Club Cultures: Music, Media, and Subcultural Capital*. Middletown, CT: Wesleyan University Press, 1996.

Tian, Xiaoli. "Face-work on Social Media in China." In *Chinese Social Media: Social, Cultural, and Political Implications*, edited by Mike Kent, Katie Ellis, and Jian Xu, 92–105. New York: Routledge, 2017.

Ting, Carol, and Famin Yi. "ICT Policy for the 'Socialist New Countryside'—A Case Study of Rural Informatization in Guangdong, China." *Telecommunications Policy* 37, no. 8 (2013): 626–638.

Tomba, Luigi. *The Government Next Door: Neighborhood Politics in Urban China*. Ithaca, NY: Cornell University Press, 2014.

Tong, Jingrong. "The Formation of an Agonistic Public Sphere: Emotions, the Internet and News Media in China." *China Information* 29, no. 3 (2015): 333–351.

Tong, Xin. "Gendered Labour Regimes: On the Organizing of Domestic Workers in Urban China." *Asian Journal of German and European Studies* 3, no. 1 (2018): 1–16.

Tong, Yanqi and Shaohua Lei. "War of Position and Microblogging in China." *Journal of Contemporary China* 22, no. 80 (2013): 292–311.

Tu, Weiming. *Confucian Thought: Selfhood as Creative Transformation*. Albany, NY: SUNY Press, 1985.

Tyler, Imogen. "Methodological Fatigue and the Politics of the Affective Turn." *Feminist Media Studies* 8, no. 1 (2008): 85–90.

Unger, Jonathan. *The Transformation of Rural China*. New York: Routledge, 2005.

Uy-Tioco, Cecilia. "Overseas Filipino Workers and Text Messaging: Reinventing Transnational Mothering." Continuum 21, no. 2 (2007): 253–265.

Van Dijck, José. *The Culture of Connectivity: A Critical History of Social Media*. New York: Oxford University Press, 2013.

Van Dijck, José, Thomas Poell, and Martijn De Waal. *The Platform Society: Public Values in a Connective World*. New York: Oxford University Press, 2018.

Vincent, Jane, and Leopoldina Fortunati, eds. *Electronic Emotion: The Mediation of Emotion via Information and Communication Technologies*. Oxford: Peter Lang, 2009.

Visser, Robin. *Cities Surround the Countryside*. Durham, NC: Duke University Press, 2010.

Vukovich, Daniel F. *Illiberal China: The Ideological Challenge of the People's Republic of China*. Singapore: Palgrave Macmillan, 2019.

Waisbord, Silvio, and Claudia Mellado. "De-westernizing Communication Studies: A Reassessment." *Communication Theory* 24, no. 4 (2014): 361-372.

Wajcman, Judy. *TechnoFeminism*. Malden, MA: Polity Press, 2004.

———. Feminist Theories of Technology. *Cambridge Journal of Economics* 34, no. 1 (2010): 143–152.

Wajcman, Judy, Michael Bittman, and Judith E. Brown. "Families without Borders: Mobile Phones, Connectedness and Work-home Divisions." *Sociology* 42, no. 4 (2008): 635–652.

Wallis, Cara. "New Media Practices in China: Youth Patterns, Processes, and Politics." *International Journal of Communication* 5 (2011): 406–436.

———. *Technomobility in China: Young Migrant Women and Mobile Phones*. New York: NYU Press, 2013a.

———. "Technology and/as Governmentality: The Production of Young Rural Women as Low-tech Laboring Subjects in China." *Communication and Critical/Cultural Studies* 10, no. 4 (2013b): 341–358.

———. "Gender and China's Online Censorship Protest Culture." *Feminist Media Studies* 15, no. 2 (2015a): 223–238.

———. "Micro-entrepreneurship, New Media Technologies, and the Reproduction and Reconfiguration of Gender in Rural China." *Chinese Journal of Communication* 8, no. 1 (2015b): 42–58.

———. "Domestic Workers and the Affective Dimensions of Communicative Empowerment." *Communication, Culture & Critique* 11, no. 2 (2018): 213–230.

———. "Domestic Workers and Performative Motherhood: WeChat, Immobile Mobility and Modernity in Beijing." In *Media in Asia: Global, Digital and Mobile*, edited by Youna Kim, 266–278. London: Routledge, 2022.

Wallis, Cara, and Anne Balsamo. "Public Interactives, Soft Power, and China's Future at and beyond the 2010 Shanghai World Expo." *Global Media and China* 1, no. 1–2 (2016): 32–48.

Wallis, Cara, and Jack Linchuan Qiu. "*Shanzhaiji* and the Transformation of the Local Mediascape in Shenzhen." In *Mapping Media in China: Region, Province, Locality*, edited by Wanning Sun and Jenny Chio, 109 125. London: Routledge, 2012.

Wang, Bin, and Catherine Driscoll. "Chinese Feminists on Social Media: Articulating Different Voices, Building Strategic Alliances." *Continuum* 33, no. 1 (2019): 1–15.

Wang, Georgette, ed. *De-Westernizing Communication Research: Altering Questions and Changing Frameworks*. New York: Routledge, 2011.

Wang, Hongwei. "Dui 'Xiao Fengya Shijian' zhong Shejiao Meiti Meijie: Lunli Wenti de Fansi" [On the Medium of Social Media in the Little Fengya Incident]. *Chuanbo yu Banquan* 4, no. 71 (2019): 191–193.

Wang, Hongzhe, and Qiu Linchuan. "Kongjian, Keji, yu Shengyin: Xin Lao Gongren Jieji Wenhua Kongjian de Bijiao Yanjiu" [Space, Technology, and Sound: A

Comparative Study of the Cultural Space of the New and Old Working Classes]. *Communication, Culture and Politics* 2 (2015): 27–60.

Wang, Jian, ed. *Shaping China's Global Imagination: Branding Nations at the World Expo*. New York: Springer, 2013.

Wang, Jing. *High Culture Fever*. Berkeley: University of California Press, 1996.

———. "The Global Reach of a New Discourse: How Far Can 'Creative Industries' Travel?" *International Journal of Cultural Studies* 7, no. 1 (2004): 9–19.

Wang, Lihua. "Neoliberalism and the Feminization of Family Survival: The Happiness Project in Four Chinese Villages." In *Social Production and Reproduction at the Interface of Public and Private Spheres*, edited by Marcia Texler Segal, Esther Ngan-ling Chow, and Vasilikie Demos, 113–138. Bingley, UK: Emerald Group Publishing, 2012.

Wang, Lin. "Dang 'Pengyouquan' zhuanbian 'Weishangjiaquan,' Weishang Fazhan Dai-lai naxie Tiaozhan" [When the "Friends Circle" Changes to the "Business Circle," What Challenges Will the Development of Micro-business Bring?]. *People's Net*. February 26, 2015. http://finance.people.com.cn

Wang, Man. "Carrying out Public Advocacy through Performance Art." *China Development Brief*. Spring 2012. https://chinadevelopmentbrief.org.

Wang, Qi. "State-society Relations and Women's Political Participation." In *Women of China: Economic and Social Transformation*, edited by Jackie West, Minghua Zhao, Xiangqun Chang, and Yuan Cheng, 19–44. Houndmills, UK: Palgrave Macmillan, 1999.

———. "Young Feminist Activists in Present-Day China: A New Feminist Generation?" *China Perspectives* no. 3 (2018a): 59–68.

———. "From 'Non-Governmental Organizing' to 'Outer-System'—Feminism and Feminist Resistance in Post-2000 China." *NORA: Nordic Journal of Women's Studies* 26, no. 4 (2018b): 260–277.

Wang, Shuaishuai, and Hongwei Bao. "'Sissy Capital' and the Governance of Non-normative Genders in China's Platform Economy." *China Information* 37, no. 3 (2023): 342–362.

Wang, Weijia, and Lijuan Yang. "'Wuyingan' yu Weibo Zhishifenzi de Dangxing" ["Wuying Case" and the Party Spirit of Intellectuals on Weibo]. *Open Times* 5 (2012): 48–62.

Wang, Xiangdong. "Nongcun Jingji Shehui Zhuangxing de Xin Moshi: Yi Shaji Dianzi Shangwu Wei Li" [A New Model for Rural Economic Transformation: A Case Study of the Shaji Model]. *Journal of Engineering Studies* 5, no. 2 (2013): 194–200.

Wang, Xiaowei. *Blockchain Chicken Farm: And Other Stories of Tech in China's Countryside*. New York: FSG Originals, 2020.

Wang, Xiling, and Liu Yiran. "Picun Xiongren Ziwo Fuquan he Weixin Gonghao de Xingxiang Jianguo" [Self-empowerment of the New Workers in Pi Village and their Self-Image Construction on WeChat Public Accounts]. *Communication & Society* 49 (2019): 75–101.

Wang, Xinyuan. *Social Media in Industrial China*. London: UCL Press, 2016.

Wang, Yini, and Judith Sandner. "Like a 'Frog in a Well'? An Ethnographic Study of Chinese Rural Women's Social Media Practices through the WeChat Platform." *Chinese Journal of Communication* 12, no. 3 (2019): 324–339.

Wang, Zheng. "Research on Women in Contemporary China." In *Guide to Women's Studies in China*, edited by Gail Hershatter, Emily Honig, Susan Mann, and Lisa Rofel, 1–43. Berkeley, CA: Institute of East Asian Studies, 1998.

———. *Finding Women in the State: A Socialist Feminist Revolution in the People's Republic of China, 1949–1964*. Berkeley: University of California Press, 2017.

———. "Feminist Struggles in a Changing China." In *Feminisms with Chinese Characteristics*, edited by Ping Zhu and Hui Faye Xiao, 117–156. Syracuse, NY: Syracuse University Press, 2021.

Wang, Zheng, and Ying Zhang. "Global Concepts, Local Practices: Chinese Feminism since the Fourth UN Conference on Women." *Feminist Studies* 36, no. 1 (2010): 40–70.

Weeks, Kathi. "Life Within and Against Work: Affective Labor, Feminist Critique, and Post-Fordist Politics." *Ephemera: Theory and Politics in Organization* 7, no. 1 (2007): 233–249.

Wei, Wei. "Street, Behavior, Art: Advocating Gender Rights and the Innovation of a Social Movement Repertoire." *Chinese Journal of Sociology* 1, no. 2 (2015): 279–304.

Weibo Corporation. Press Release, November 14, 2019. https://weibocorporation.gcs-web.com.

———. "Weibo Reports First Quarter 2023 Unaudited Financial Results and Dividend." May 25, 2023. http://ir.weibo.com.

"Weibo Reaches 100 Million Daily Users." eMarketer. January 12, 2016. www.emarketer.com.

Weidman, Amanda. "Anthropology and Voice." *Annual Review of Anthropology* 43 (2014): 37–51.

Welland, Sasha Su-Ling. *Experimental Beijing: Gender and Globalization in Chinese Contemporary Art*. Durham, NC: Duke University Press, 2018.

Wen, Jiabao. "New Socialist Countryside—What Does it Mean?" *Beijing Review* 14 (April 6, 2006). www.bjreview.com.cn.

Wielander, Gerda. "Chinese Happiness: A Shared Discursive Terrain." In *Chinese Discourses on Happiness*, edited by Gerda Wielander and Derek Hird, 1–24. Hong Kong: Hong Kong University Press, 2018.

Wines, Michael, and Sharon LaFraniere. "In Baring Facts of Train Crash, Blogs Erode China Censorship." *New York Times*, July 28, 2011.

Wise, J. Macgregor. "Assemblage." In *Gilles Deleuze: Key Concepts*, edited by Charles J. Stivale, 77–87. Montreal: McGill-Queen's University Press, 2005.

Women's Media Monitor Network (Funü Chuanmei Jiance Wangluo). *Ting, Bieju Nüsheng: Nüsheng Dianzibao Liangnian Xuan* [Listen, Unique Women's Voice: Selections of Two Years of Women's Voice E-paper]. 2009–2011.

Wong, Winnie. *Van Gogh on Demand: China and the Readymade*. Chicago: University of Chicago Press, 2014.

Wu, Angela Xiao, and Yige Dong. "What is Made-in-China Feminism(s)? Gender Discontent and Class Friction in Post-Socialist China." *Critical Asian Studies* 51, no. 4 (2019): 471–492.

Wu, Changchang. "Micro-blog and the Speech Act of China's Middle Class: The 7.23 Train Accident Case." *Javnost - The Public* 19, no. 2 (2012): 43–62.

Wu, Di. *Affective Encounters: Everyday Life among Chinese Migrants in Zambia*. New York: Routledge, 2020.

Wu, Guoguang, and Helen Lansdowne. "Introduction." In *Gender Dynamics, Feminist Activism and Social Transformation in China*, edited by Guoguang Wu, Feng Yuan, and Helen Lansdowne, 1–11. New York: Routledge, 2019.

Wu, Yenna. "Making Sense in Chinese 'Feminism'/Women's Studies." In *Dialogue and Difference: Feminisms Challenge Globalization*, edited by Marguerite Waller and Sylvia Marcos, 29–52. New York: Palgrave Macmillan, 2005.

Xi, Jinping. *Xi Jinping Guanyu Shixian Zhonghua Minzu Weida Fuxing de Zhongguo Meng: Lunshu Gaobian* [Xi Jinping on Realizing the China Dream's Great Rejuvenation of the Chinese Nation: Discussion Edition]. Beijing: Zhongyang Wenxian Chubanshe, 2013.

———. "Zai Wenyi Gongzuo Zuotan Huishangde Jianghua" [Talks at the Forum on Literature and Art]. Xinhua. October 15, 2014. www.xinhuanet.com.

———. "Zhashi Tuidong Gongtong Fuyu" [Solidly Promote Common Prosperity]. *Seeking Truth*, October 15, 2021. www.qstheory.cn.

Xia, Jun. "Linking ICTs to Rural Development: China's Rural Information Policy." *Government Information Quarterly* 27, no. 2 (2010): 187–195.

Xiang, Hardy Yong, and Patricia Ann Walker, eds. *China Cultural and Creative Industries Reports*. New York: Springer Science & Business Media, 2013.

Xiao, Josh. "China Tells Women to Respect 'Family Values' in Revised Law." *Bloomberg News*, November 3, 2022. www.bloomberg.com.

Xinhua. "Zhonghua Renmin Gongheguo Guomin Jingji he Shehui Fazhan Dishisange Wunian Guihua Gangyao" [Outline of the 13th Five-Year Plan for National Economic and Social Development of the People's Republic of China]. March 17, 2016. www.gov.cn.

———. "Full Text of Xi Jinping's Report at the 19th CPC National Congress." November 3, 2017. www.xinhuanet.com.

———. "Zhonggong Zhongyang Guowuyuan Yinfa 'Xiangcun Zhenxing Zhanlüe Guohua (2018–2022 nian)'" [The Central Committee of the Communist Party of China and the State Council Issued the "Strategic Plan for Rural Revitalization (2018–2022)"]. September 26, 2018. www.xinhuanet.com.

———. "China Focus: Xi Declares China a Moderately Prosperous Society in All Respects." July 1, 2021. www.xinhuanet.com.

Xiong, Jing. "From Margin to Centre: Feminist Mobilizations in Digital China." In *Gender Dynamics, Feminist Activism and Social Transformation in China*, edited by Guoguang Wu, Feng Yuan, and Helen Lansdowne, 213–232. New York: Routledge, 2019.

Xiong, Jing, and Dušica Ristivojević. "#MeToo in China: How Do the Voiceless Rise Up in an Authoritarian State?" *Politics & Gender* 17, no. 3 (2021): 490–499.

Xue, Aviva Wei, and Kate Rose. *Weibo Feminism: Expression, Activism, and Social Media in China*. London: Bloomsbury Academic, 2022.

Yan, Hairong. *New Masters, New Servants: Migration, Development, and Women Workers in China*. Durham, NC: Duke University Press, 2008.

Yan, Yunxiang. *The Flow of Gifts: Reciprocity and Social Networks in a Chinese Village*. Stanford, CA: Stanford University Press, 1996.

———. *The Individualization of Chinese Society*. New York: Berg, 2009.

———. "The Changing Moral Landscape." In *Deep China: The Moral Life of the Person*, by Arthur Kleinman, Yunxiang Yan, Jing Jun, Sing Lee, Everett Zhang, Pan Tianshu, Wu Fei, and Jinhua Guo, 36–77. Berkeley: University of California Press, 2011.

———. "The Drive for Success and the Ethics of the Striving Individual." In *Ordinary Ethics in China*, edited by Charles Stafford, 263–292. New York: Routledge, 2013.

———. "Doing Personhood in Chinese Culture: The Desiring Individual, Moralist Self and Relational Person." *The Cambridge Journal of Anthropology* 35, no. 2 (2017): 1–17.

———. "The Politics of Moral Crisis in Contemporary China." *The China Journal* 85, no. 1 (2021): 96–120.

Yang, Fan. *Faked in China: Nation Branding, Counterfeit Culture, and Globalization*. Bloomington: Indiana University Press, 2016.

Yang, Fenggang. *Religion in China: Survival and Revival under Communist Rule*. New York: Oxford University Press, 2011.

Yang, Guobin. *The Power of the Internet in China: Citizen Activism Online*. New York: Columbia University Press, 2009.

———. "Contesting Food Safety in the Chinese Media: Between Hegemony and Counter-hegemony." *The China Quarterly* 214 (2013): 337–355.

———. "Demobilizing the Emotions of Online Activism in China: A Civilizing Process." *International Journal of Communication* 12 (2018): 1945–1965.

Yang, Jie. "'Fake Happiness': Counseling, Potentiality, and Psycho-Politics in China." *Ethos* 41, no. 3 (2013): 292–312.

———. "The Happiness of the Marginalized: Affect, Counseling and Self-Reflexivity in China." In *The Political Economy of Affect and Emotion in East Asia*, edited by Jie Yang, 45–61. New York: Routledge, 2014a.

———. "The Politics of Affect and Emotion: Imagination, Potentiality and Anticipation in East Asia." In *The Political Economy of Affect and Emotion in East Asia*, edited by Jie Yang, 3–28. New York: Routledge 2014b.

———. *Unknotting the Heart: Unemployment and Therapeutic Governance in China*. Ithaca, NY: Cornell University Press, 2015.

———. "The Politics and Regulation of Anger in Urban China." *Culture, Medicine, and Psychiatry* 40 (2016): 100–123.

———. "'Bureaucratic *Shiyuzheng*': Silence, Affect, and the Politics of Voice in China." *HAU: Journal of Ethnographic Theory* 11, no. 3 (2021): 972–985.

Yang, Mayfair Mei-hui. *Gifts, Favors, and Banquets: The Art of Social Relationships in China*. Ithaca, NY: Cornell University Press, 1994.

———. "From Gender Erasure to Gender Difference: State Feminism, Consumer Sexuality, and Women's Public Sphere in China." In *Spaces of Their Own: Women's Public Sphere in Transnational China*, edited by Mayfair Mei-hui Yang, 35–67. Minneapolis: University of Minnesota Press, 1999.

———. "The Resilience of *Guanxi* and its New Deployments: A Critique of some New *Guanxi* Scholarship." *The China Quarterly* 170 (2002): 459–476.

Yang, Peidong, and Tang Lijun. "'Positive Energy': Hegemonic Intervention and Online Media Discourse in China's Xi Jinping Era." *China: An International Journal* 16, no. 1 (2018): 1–22.

Yang, Yue. "When Positive Energy meets Satirical Feminist Backfire: Hashtag Activism during the COVID-19 Outbreak in China." *Global Media and China* 7, no. 1 (2022): 99–119.

Ye, Jingzhong. "Left-behind Children: The Social Price of China's Economic Boom." *Journal of Peasant Studies* 38, no. 3 (2011): 613–650.

Yen, Dorothy A., Bradley R. Barnes, and Cheng Lu Wang. "The Measurement of Guanxi: Introducing the GRX Scale." *Industrial Marketing Management* 40, no. 1 (2011): 97–108.

Yi, Lin. "Individuality, Subjectivation, and their Civic Significance in Contemporary China: The Cultivation of an Ethical Self in a Cultural Community." *China Information* 33, no. 3 (2019): 329–349.

Yin, Siyuan. "Producing Gendered Migration Narratives in China: A Case Study of *Dagongmei Tongxun* by a Local NGO." *International Journal of Communication* 10 (2016): 4304–4323.

———. "Cultural Production in the Working-Class Resistance: Labour Activism, Gender Politics, and Solidarities." *Cultural Studies* 34, no. 3 (2020): 418–441.

———. "Re-Articulating Feminisms: A Theoretical Critique of Feminist Struggles and Discourse in Historical and Contemporary China." *Cultural Studies* 36, no. 6 (2022): 981–1004.

Yin, Siyuan, and Yu Sun. "Intersectional Digital Feminism: Assessing the Participation Politics and Impact of the MeToo Movement in China." *Feminist Media Studies* 21, no. 7 (2021): 1176–1192.

You, Shibing, Jingru Ren, and Pei Zhang. "Yi Nongcun Dianshang Bu 'Sannong' Duanban" (Use Rural E-commerce to Make up for the "Three Rural" Shortcomings). *People's Daily*, January 19, 2017. http://theory.people.com.cn.

Yu, Haiqing, and Lili Cui. "China's E-Commerce: Empowering Rural Women?" *The China Quarterly* 238 (2019): 1–20.

Yu, Xueyi. "Why I Quit Alibaba's Big Push toward Countryside Commerce." Sixth Tone. March 9, 2017. www.sixthtone.com.

Yuan, Elaine J. *The Web of Meaning: The Internet in a Changing Chinese Society*. Toronto: University of Toronto Press, 2021.

Zeng, Jing. "#MeToo as Connective Action: A Study of the Anti-Sexual Violence and Anti-Sexual Harassment Campaign on Chinese Social Media in 2018." *Journalism Practice* 14, no. 2 (2020): 171–190.

Zeng, Jinyan. "The Politics of Emotion in Grassroots Feminist Protests: A Case Study of Xiaoming Ai's Nude Breasts Photography Protest Online." *Georgetown Journal of International Affairs* 15, no. 1 (2014): 41–52.

Zeng, Yiwu, and Hongdong Guo. "Nongchanpin Taobaocun Xingcheng Jili: Yige Duo-anli Yanjiu" [The Formation Mechanism of Agro-Taobao Villages: A Multiple-Case Study]. *Agricultural Economy* 4 (2016): 39–48.

Zenglein, Max J., and Anna Holzmann. "Evolving Made in China 2025: China's Industrial Policy in the Quest for Global Tech Leadership." Mercator Institute for China Studies No. 8, July 2019. https://kritisches-netzwerk.de.

Zhang, Everett. "Introduction: Governmentality in China." In *Governance of Life in Chinese Moral Experience: The Quest for an Adequate Life*, edited by Everett Zhang, Arthur Kleinman, and Tu Weiming, 1–30. New York: Routledge, 2011.

Zhang, Guannan. "Zhonghua Tianyuannüquan Ci Yikao: Wangluo Huanjing Chuanbozhong Nüxingquanli de Wuminghua" [Examining the Meaning of Chinese Countryside Feminism: The Stigmatization of Feminism in the Virtual World]. *Media Observer* 7 (2018): 99–101.

Zhang, Hui. "The Ethics of Envy Avoidance in Contemporary China." In *Ordinary Ethics in China*, edited by Charles Stafford, 115–132. London: Bloomsbury, 2013.

Zhang, Hui. "Nanny Industry Criticized after Family Murdered in own Home." *People's Daily*, June 29, 2017. http://en.people.cn.

Zhang, Jingyuan. *Dangdai Nüxingzhuyi Wenxue Piping* [Contemporary Feminist Literary Criticism]. Edited and translated by Jingyuan Zhang. Beijing: Beijing University Press, 1992.

Zhang, Jinman. "Feminist Responses to COVID-19 in China through the Lens of Affect." *Feminist Media Studies* 23, no. 4 (2023): 1327–1343.

Zhang, Li. *Strangers in the City: Reconfigurations of Space, Power, and Social Networks within China's Floating Population*. Stanford, CA: Stanford University Press, 2001.

———. "Cultivating Happiness: Psychotherapy, Spirituality, and Well-Being in a Transforming Urban China." In *Handbook of Religion and the Asian City*, edited by Peter van der Veer, 315–332. Berkeley: University of California Press, 2015.

———. "The Rise of Therapeutic Governing in Postsocialist China." *Medical Anthropology* 36, no. 1 (2017): 6–18.

———. *Anxious China: Inner Revolution and Politics of Psychotherapy*. Berkeley: University of California Press, 2020.

Zhang, Lin. *The Labor of Reinvention: Entrepreneurship in the New Chinese Digital Economy*. New York: Columbia University Press, 2023.

Zhang, Nan. "Rural Women Migrant Returnees in Contemporary China." *Journal of Peasant Studies* 40, no. 1 (2013): 171–188.

Zhang, Pengyi. "Social Inclusion or Exclusion? When *Weibo* (Microblogging) Meets the New Generation of Rural Migrant Workers." *Library Trends* 62, no. 1 (2013): 63–80.

Zhang, Su-ling. "Wan Qing zhi Wusi Shiqi Zhishifenzi de Xingbiehua yu Jiqi Shehui Wenhua Yiyun" [The Gender Discourse of Intellectuals and its Sociocultural Meaning from Late-Qing Dynasty to the May 4th Movement]. *Collection of Women's Studies* 3 (2007): 26–32.

Zhang, Weiyu. *The Internet and New Social Formation in China: Fandom Publics in the Making.* New York: Routledge, 2016.

Zhang, Yanhua. *Transforming Emotions with Chinese Medicine: An Ethnographic Account from Contemporary China.* Albany, NY: SUNY Press, 2007.

———. "Cultivating Capacity for Happiness as a Confucian Project in Contemporary China: Texts, Embodiment, and Moral Affects." In *Chinese Discourses on Happiness*, edited by Gerda Wielander and Derek Hird, 150–168. Hong Kong: Hong Kong University Press, 2018.

Zhao, Jinqiu. *The Internet and Rural Development in China: The Socio-structural Paradigm.* Bern: Peter Lang, 2008.

Zhao, Yuezhi. *Communication in China: Political Economy, Power, and Conflict.* Lanham, MD: Rowman & Littlefield, 2008.

Zhu, Ping, and Hui Faye Xiao. "Feminisms with Chinese Characteristics: *An Introduction*." In *Feminisms with Chinese Characteristics*, edited by Ping Zhu and Hui Faye Xiao, 1–34. Syracuse, NY: Syracuse University Press, 2021.

Zi, Li. *Online Urbanization: Online Services in China's Rural Transformation.* Singapore: Springer, 2019.

Zigon, Jarrett. "An Ethics of Dwelling and a Politics of World-building: A Critical Response to Ordinary Ethics." *Journal of the Royal Anthropological Institute* 20, no. 4 (2014): 746–764.

INDEX

Page numbers in italics indicate Photos.

ABOUT THE AUTHOR

CARA WALLIS is an Associate Professor in the Department of Communication and Media at the University of Michigan. She is the author of *Technomobility in China: Young Migrant Women and Mobile Phones* (NYU Press, 2013), which won the 2013 James W. Carey Media Research Award and the 2014 Bonnie Ritter Award.

www.ingramcontent.com/pod-product-compliance
Lightning Source LLC
Chambersburg PA
CBHW031142020426
42333CB00013B/488